20-50,000 / week

Construction Technology
Today and Tomorrow

Construction
Technology
Today and Tomorrow

James F. Fales, Ed.D., CMfgE
Professor and Chairman
Department of Industrial Technology
Ohio University
Athens, Ohio

GLENCOE
Macmillan/McGraw-Hill

Lake Forest, Illinois
Columbus, Ohio
Mission Hills, California
Peoria, Illinois

Send all inquiries to:
Glencoe Division, Macmillan/McGraw-Hill
809 West Detweiller Drive
Peoria, IL 61615-2190

ISBN 0-02-675754-0 (Student Text)
ISBN 0-02-675755-9 (Teacher's Resource Guide)
ISBN 0-02-675756-7 (Student Workbook)

2 3 4 5 6 7 8 9 10   99 98 97 96 95 94 93 92 91

ISBN 0-02-675758-3 (Teacher's Annotated Edition)

||| ACKNOWLEDGMENTS

The publisher hereby gratefully acknowledges the cooperation and assistance received from many persons and companies during the development of *Construction Technology: Today and Tomorrow*. Special recognition is given to the following persons for their contributions:

Dave Pullias
Richardson Independent School District
Richardson, TX

Ken Smith
Technology Education Resource Teacher
Charles County Public Schools
La Plata, MD

James D. Stemple
Technology Education Teacher
Stafford Senior High School
Stafford, VA

Larry L. Stiggins
Technology Education Instructor
J.T. Hutchinson Junior High School
Lubbock, TX

Chapter number photo: Ping Amranand, Sylvania horse barn, Meadowfarm Stable, Orange, VA

Cover photo by Digital Art/Westlight

||| TABLE OF CONTENTS

INTRODUCTION

The construction industry today is an exciting and expanding field. When you think of construction, you may think of bulldozers and high-rise buildings. The world of construction involves much more, however. Construction involves not only the building of homes and office buildings. It is also concerned with the building of highways, tunnels, airports, and dams. It also involves more people than the construction workers you see on construction sites. Many others are working behind the scenes. These people are architects, engineers, and office workers, among others. Construction today is a large and growing industry. It builds on the technical accomplishments of the past. It also depends on more recent technological developments — inventions such as the laser and the computer.

Construction Technology: Today and Tomorrow provides a complete survey of the exciting construction industry. It allows you to explore all areas of this rapidly-growing field, including the most recent changes in technology. To give you some idea of the design of this text, let's discuss its organization. First, we'll discuss the photo-essay that opens the text. Then we'll discuss the table of contents and the various features in each of the text chapters.

THE ORGANIZATION OF THIS BOOK

Photo-Essay

Placed immediately before Section I, the photo-essay is the first part of this book. Titled "The Development of Construction Technology," this photo-essay provides a window on the exciting world of construction technology. With photos and text, it provides a general overview of the development of construction technology throughout history. In reading this photo-essay, look closely at the photos that accompany the text. They have been carefully selected to highlight the information in the text. Each photo carries a caption, or a description of what is shown in the photo. Read these captions carefully. They go beyond simply describing what is shown in the photo. They provide other valuable information that relates to the topics discussed in the text. When you have finished reading the photo-essay, you will have a better idea of the subjects that will be discussed in this book. You also will gain a greater knowledge of the historical developments that have led to the growth of construction technology.

Table of Contents

If you turn to the table of contents, you will see that this textbook has sixteen chapters. Each of these chapters discusses a separate topic in the field of construction technology. You also will notice that the chapters are grouped into sections. All of the chapters in a particular section relate to the basic theme of that section. Look, for example, at Section II ("Materials, Tools, and Processes"). You will see that each of the chapters in that section deals with the section topic.

Chapter Features

The features in each chapter are designed to help you learn the information presented there. Each chapter contains a variety of helpful and interesting features. These features will allow you to gain more information from this book. Each feature presents additional information relating to the topic being discussed. The types of features appearing in each chapter are discussed below. Watch for them as you read the chapters of this book.

- *Learning Objectives.* A short list of learning objectives appears at the beginning of each chapter. These learning objectives provide goals that you should set for yourself before you begin to read and study the information in the chapter. Then, as you read each chapter, you should occasionally refer back to the chapter objectives. By doing this, you will be able to make sure that you still have them in mind. These learning objectives are important. They relate to the key ideas in the chapter.

- *Terms to Know.* The Terms to Know also are listed at the beginning of each chapter. Each of these terms is set in **boldfaced type** within the chapter. Pay special attention to these terms. They signal important information. Each of the boldfaced terms in the chapter is also defined in the glossary at the back of the book.

- *Did You Know?* These short paragraphs offer information that builds on the information presented in the text. Usually, this new information relates construction technology to other subjects you may be studying. These subjects might be history, geography, or mathematics. All of these short features are designed to give you more information on construction technology.

- *Construction Facts.* Each chapter contains one long feature. This feature is usually toward the end of the chapter, just before the chapter review material. Each of these features highlights an important event in the history of construction technology. Each feature is illustrated with a photo or drawing chosen to provide more information on the subject.

- *For Discussion.* As you read the text, you may have some questions on the information. You might also want to share some of your ideas with the class. The short features titled "For Discussion?" will allow you to share your ideas with others. Each of these features presents a question relating to information in the text. These questions will provide a springboard for class discussion of an interesting topic.

- *Health and Safety.* Safe work habits are important in the construction industry. These features highlight health and safety practices and offer tips on working safely. They also highlight important events in the development of safe work practices.

End-of-Chapter Review Material

When you have finished reading each chapter, you will come to the end-of-chapter review material. This material consists of a chapter summary, chapter review questions, and activities. All of these are designed to help you and your teacher judge how well you have learned the information presented in the chapter. Remember the learning objectives at the beginning of the chapter? All of the material at the end of the chapter relates to these learning objectives. Let's look at each of the items in the end-of-chapter material.

- *Chapter Summary.* A summary is a statement of key points. The chapter summary is a statement, then, of the key points in the chapter. By reading the chapter summary, you can review the main points of information discussed in the chapter. You will want to read the chapter summary before you look at the review questions.

- *Test Your Knowledge.* The review questions at the end of each chapter are titled "Test Your Knowledge." They are designed to call your attention to the key points of information in the chapter. Your ability to answer these questions correctly will help you judge how well you have learned the information in the chapter.

- *Activities.* Each chapter closes with a set of short activities. Some of these activities are designed to develop skills in areas that relate to some of the other courses you may be taking. For most of these activities, you will not need anything other than paper and a pencil. These activities follow the "Test Your Knowledge" questions.

Each of these end-of-chapter activities offers you an opportunity to develop a certain skill. The skills you can develop are in the areas of math, science, social studies, and communication. Each end-of-chapter activity is marked with one of the symbols shown below. These symbols identify the type of skill that can be developed by the activity.

Math   **Social Studies**

Science   **Communications**

Section and End-of-Section Activities

As mentioned above, the chapters of this book are grouped into sections. Each section deals with a specific theme in construction technology. A list of all the chapters in the section appears at the beginning of the section. These chapters are listed by number and title.

- *End-of-Section Activities.* A set of activities follows the last chapter in each section. These activities are longer than the activities that are included at the end of each chapter. These end-of-section activities also usually require materials other than paper and a pencil. They sometimes will provide you with an opportunity to build something. Also, they often present you with a chance to demonstrate teamwork. In some of these activities, you will be working in a two-person team or in a small group. Each of these activities is designed to make it easy for you to follow the directions. For every activity there is an objective, a list of the materials needed, and a numbered list of the steps of procedure.

Glossary and Index

- *Glossary.* Remember the boldfaced terms in the chapters? Each of these terms is listed

also in the glossary at the end of the book. A glossary is a list of terms explained. The glossary is alphabetized. If you need to know the meaning of a boldfaced term used in the text, look in the glossary. You will find its definition there.

- *Index.* The index is an important part of any book. Often, the value of a good index is not properly appreciated. The word *index* comes from the Latin word for "to indicate." An index entry, then, points you toward something. It helps you learn more about a term. If you would like to know where a certain subject is discussed in this book, look in the index. It will give you the numbers of the pages that provide information on that subject.

HEAD WEIGHTS

Turn forward a few pages to Chapter 1. You will notice that each major part of that chapter is introduced by a short title. These titles are set in different type sizes. The size of the type indicates the importance of the subject being discussed. These short titles are called "heads." There are heads in all of the chapters in this book.

The heads in a chapter can serve as stepping-stones in helping you gain information. They alert you to what will be covered in that part of the chapter. Before you read a chapter, scan the heads that introduce the various parts of the chapter. Scanning these heads will give you a general idea of the topics that will be discussed in the chapter.

Regardless of your career goals, you will find that this book is a helpful guide to one of the key technologies in the modern world — construction technology. By gaining a general knowledge of the materials and processes of this growing industry, you will have a better understanding of our modern and changing world.

THE DEVELOPMENT OF CONSTRUCTION TECHNOLOGY

TOOLS AND TECHNOLOGY

Technology is the use of technical methods to obtain practical results. Technology as it relates to construction is known as **construction technology**. The instruments of technology are tools. A tool is anything that makes a job easier. For example, a laser-guided bulldozer is a tool, just as is a common shovel. Because our age has developed complex tools, it has been called the Age of Technology.

The technology we use today builds on developments from throughout the entire experience of the human race. Our own time is the present stage of history, a word that comes from the Greek word for "to know." History, then, is the time about which we know. It stretches back from the present day to the end of that period known as prehistory. Prehistory is that time about which we know little. Much of human history and prehistory is concerned with the slow, but gradual, development of the technological skills needed for survival. Because shelter is a basic need, many of these skills related to construction.

Let's take a brief look at the history of construction technology. This will help give us a better appreciation of the construction developments of our own time.

SKILLS TO MEET BASIC NEEDS

Shelter, along with food and clothing, was one of the three main concerns of early humans.

A laser is a beam of intense light that can be accurately focused. The blade of this grader is guided by a laser. A sensing device on the blade follows the laser beam. The beam, then, guides the blade of the grader. It ensures that the blade cuts to a certain depth. The laser is a fairly new device. Its use in construction is just one example of the way in which the construction industry makes use of available technology.

The first shelters were natural—cave entrances and rock overhangs. These were not built or even greatly adapted for human use. They were merely found and used for what they were—protection from the weather.

As long as people camped in one place, the same shelter could be used day after day. However, as people moved in a ceaseless search for food, a new shelter would have been needed. If a natural shelter could not be found, a shelter would have to be built. From necessity, people developed basic construction skills.

These first shelters actually built by humans were made of available materials—tree branches and stones. These materials would have been used just as they were found. Early humans lacked the tools to greatly alter materials to suit their purposes. Though we might find them crude, these simple shelters would have stretched to the limit the technological skills of early humans.

CONSTRUCTION TECHNOLOGY AND THE DEVELOPMENT OF CITY LIFE

At some point, people realized that it would be easier to carry a portable shelter rather than to build a new shelter at each campsite. To be portable, these shelters needed also to be light.

This dwelling of the Algonkian Amerindians was portable. It could be quickly taken down and transported to another place. It provided a reliable shelter.

Generally, they consisted of little more than a flexible covering, such as animal skins, stretched over an arrangement of poles. Still, as long as humans traveled from place to place, they were limited in their design of shelters. This limitation hampered their development of new construction skills.

The products of construction technology reflect the needs of society. In this view of a city block, you see several different types of buildings. Each type of building is used for a different purpose. Some are used as workplaces. Others are used as apartments. Others are used for recreation. The buildings shown here reflect the complexity of our modern social life. Today, we use certain buildings for certain purposes. In ancient times, most human activities were carried on in one place. The need for buildings for different purposes has proved a challenge to modern construction technology. Working with engineers, architects design buildings for a wide range of uses.

THE SPECIALIZATION OF SKILLS

When people learned to grow plants for food, they no longer needed to seek out a new food source each day. This development—the beginning of agriculture—led to the founding of human settlements. Settled in one place, people began to practice those construction skills that would ensure their survival in that place. They built fortifications and dug wells.

To earn their living, people began to concentrate on doing one thing well. For example, if they were farmers, they became better at growing crops. In short, people became more specialized in their job skills. Becoming more specialized, they also became dependent on others for the things they could not provide for themselves. To gain animals, food crops, and needed handmade items, they began to trade with others.

CONSTRUCTION TECHNOLOGY AND THE GROWTH OF TRADE

In this gradual development of trade there lay the seeds of a growing commerce. Commerce is the buying and selling of goods that require transportation from one place to another. The development of commerce was a major spur to the growth of construction technology. Commerce became important to the economy of these societies. The word economy relates to the production and sale of goods and services.

The economic survival of these early town-dwellers depended on a thriving commerce. To create and maintain this, they needed to develop the technical skills to build roads and bridges.

These were needed to establish trade routes, to link one town with the next. By making commerce easier, these transportation links helped ensure the survival of city life.

Some of these small towns prospered, becoming trading centers. In them, buildings of a more permanent nature began to be constructed. Again, the materials used would have been those at hand—stone and wood in Europe, clay in the Middle East. Now, though, the material was being shaped. Stone was being cut into building blocks. Tree trunks were being shaped into building beams. Clay was being formed into sun-dried bricks.

To meet these construction needs, there began to emerge a group of workers who gained most of their livelihood from the practice of a single construction skill. The skill may have been carpentry or stonemasonry. Such specialization allowed workers to set work standards. Using their skills, these workers learned to shape their environment, or surroundings, in new ways. They began to create a **built environment**. In time, changing slowly over a period of thousands of years, this built environment would develop into an environment recognizable to all of us. It would be the city.

Until fairly recently, humans spent a great deal of their energy fortifying the places in which they lived. Security was one of their principal considerations. Some of the largest early construction projects were fortifications. This Iron Age hill fort in Dorset, England is an example of such a fortification. The fort is known as Maiden Castle. The earth used to build Maiden Castle was moved by hand over a period of many years. The people who built the fort lived around it and inside it. In 70 A.D., the invading Romans conquered the people living here. The survivors were moved to a site 2 miles [3.2 km] away. There, a new town—visible in the distance—developed.

THE REBIRTH OF CONSTRUCTION TECHNOLOGY

As can be seen, a well-developed and vigorous city life was essential to the development of construction technology. In no period of human history was this shown more clearly than during the so-called Dark Ages. The Dark Ages is the name for the period from the fall of Rome in 476 to the rebirth of city life in about 1000. During this period of about 500 years, construction dropped off. Many of the roads and buildings of the Romans—who were master builders—were dismantled. Their stones were then used for other purposes, usually agricultural. City life diminished. People returned to working the land.

In about 1000, city life began to revive. Once again, there was a growth in the development of skilled trades, especially building trades. Commerce began to thrive. Better roads were needed, as were more bridges and more functional build-

The Romans were the first people to emphasize the importance of good roads. They were masters at roadbuilding. Used mainly to move troops rapidly throughout the empire, many of their roads were divided into slow and fast lanes. Also, their roads usually were built to follow a straight line. They went over natural obstacles, such as hills, rather than around them. Modern interstate highways employ both of these principles. They allow two lanes of traffic—fast and slow—in each direction. Whenever possible, interstates are laid out to run in a straight line—a straight line being the shortest distance between two points.

ings. Carpenters and stonemasons, as well as other skilled workers, organized themselves into groups called guilds. These set standards of workmanship. They also ensured that their members worked at an acceptable level of skill. The cathedrals of Europe—most of them built between 1100 and 1300—are the surviving momuments to the skills of these medieval carpenters and masons.

THE BUILT ENVIRONMENT

Just as the the prosperous growth of city life fostered the growth of construction technology, it also created problems for it to solve. As an example, let us look at London. In the late nineteenth century, that city became enormously overcrowded. This overcrowding caused serious transportation problems. City streets were crowded. The city had no efficient public transportation system. The roads into the city also were jammed with traffic. The transportation problem within the city was solved by the con-

The building of the medieval European cathedrals was an important point in the development of construction technology. Large teams of laborers and skilled workers were needed to complete these projects. Keeping track of materials and work schedules was a difficult task, requiring smooth organization. The task was not eased by the fact that construction was often carried on over a period of forty or fifty years. These cathedrals represent one of the most organized and sustained building efforts of the Middle Ages. This shows Notre Dame in Paris.

struction of the Tube. This subway was the world's first underground transportation system. Transportation into and out of London was greatly improved through the construction of a number of large railroad stations. Each of these construction projects involved the building of several different types of structures. Tunnels, bridges, overpasses, and roadways were all built as part of these projects.

In the United States, many of the outstanding construction projects of the nineteenth century also involved building transportation links. Two of the greatest marvels of construction were the transcontinental railroad and the Brooklyn Bridge (see page 263). Each of these massive construction projects drew on the latest construction technology of the day.

‖‖ CONSTRUCTION TECHNOLOGY: TODAY AND TOMORROW

Some think of construction technology as relating only to buildings. However, it relates also to the building of highways and bridges as well as to the building of tunnels and airport runways. We travel to school or to work on roads and highways. These may pass through tunnels or across bridges. All of these—as well as the buildings in which we live, go to school, and work—are examples of the uses of construction technology. As you can see, the works of construction technology are essential in modern communication and transportation systems.

In our own time, we have used construction technology to build an efficient network of interstate highways, rail systems, and airports. This has freed us from the need to build our structures from the materials at hand. Look around you and you will find that the materials used to construct the buildings in which you live and go to school were probably brought from another place. As materials have become more readily available, builders have been able to concentrate more on their use. This has led to the use of construction materials that save time. Of course, these materials can also cut costs and provide greater strength and safety. New high-strength steels have enabled us to build structures that would have been impossible before. Plas.ics have found an increasing use in construction. New instruments, such as lasers, also are being used in construction.

In our own century, one of the most spectacular construction projects has been the Verrazano-Narrows Bridge. With a length of 4,260 feet (1,298 m), it spans New York Harbor to connect Staten Island with Brooklyn. Its suspension cables weigh nearly 10,000 tons each. Perhaps the most famous of the New York bridges is the Brooklyn Bridge. For more information on the Brooklyn Bridge, refer to page 263.

Modern transportation systems allow builders a wide choice of construction materials. On the American frontier, the log cabin was the principal dwelling in the midwest. The sod house was built in the Plains states of Nebraska and North and South Dakota. Each of these dwellings, like the dwellings of humans long ago, was built using available resources. Today, an expanded construction technology and an effective transportation system allow builders to choose from a wide range of building materials. Modern skyscrapers are built from a variety of materials.

These new developments in construction technology are encouraging. They have given us the confidence and ability to undertake massive construction schemes. One of these is the Chunnel, the undersea link between England and France (see page 235).

Projects such as the Chunnel require careful organization. Each stage of construction must be carefully coordinated with the others. Computers will be essential in managing such large construction projects. Such great projects will once again test the limits of our construction skills. In testing these skills, we are again working against the same limits imposed on the first people who sought to build a simple shelter from the materials at hand. Once again, we will be testing our skills against the circumstances. If we meet the challenge, we will have expanded the frontiers of construction technology. We will have met new goals. As before, these goals will encourage us to work still harder to focus the uses of construction technology to meet our needs.

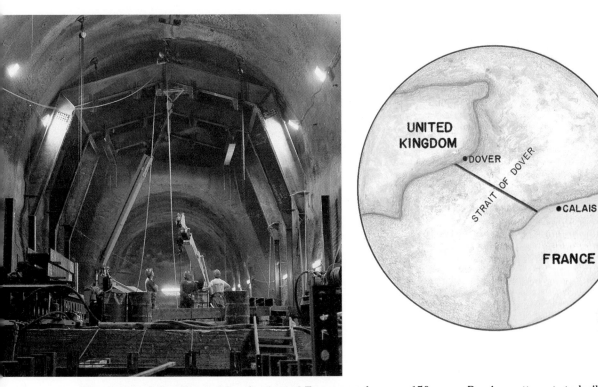

The dream of the Chunnel has fascinated Europeans for over 150 years. Previous attempts to build such a tunnel beneath the English Channel have failed. Now, the technology and financing are available. Construction has now begun on what some see as the construction project of the century. For more information on this exciting project, refer to page 235. The map on the right shows the route of the undersea Chunnel between the United Kingdom and France.

A computer is hardware. The programs that are used on a computer are known as software. A number of computer programs are especially designed to assist builders in managing construction projects. For example, computers can be used in ordering materials and scheduling construction activities. They also can be used to prepare financial projects and assess the profit and loss on a construction project. In the management of a large construction project, the computer can be a valuable tool. This shows construction drawings created on a computer.

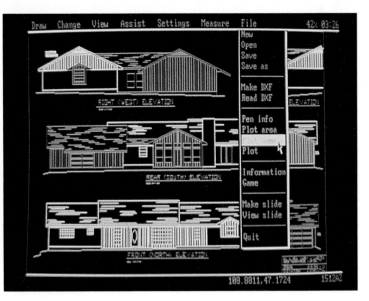

SECTION

I

INTRODUCTION

CHAPTER

1

INTRODUCTION TO CONSTRUCTION

Terms to Know

construction process
construction
    technology
financing

general contractor
scheduling
transfer of
    ownership

Objectives

When you have completed reading this chapter, you should be able to do the following:

- Describe construction technology.
- Identify the effects of construction in our world.
- Explain the basic procedure for constructing a building.
- Explain why a construction project should be considered carefully before it is begun.

The construction industry as we know it is the result of thousands of years of development. Today we have heavy equipment and sophisticated tools that help us construct tall buildings and other large structures within a short period of time. Fig. 1-1.

The tools and equipment, as well as the processes construction workers use to build structures, are the result of construction technology. **Construction technology** can be described as our use of tools, materials, and processes to build structures such as buildings, highways, and dams. To say this another way, construction technology refers to all the knowledge we have gained about how to build structures to meet our needs.

BUILDINGS AND MUCH MORE

When we think of construction, we generally think first of buildings. However, construction includes every type of structure that people build. Construction plays such a universal role in our lives because almost every type of need can be at least partially satisfied by some type of construction.

Community Development

Construction plays an important part in shaping the communities in which we live. Most communities have homes, office buildings, churches, factories, banks, stores, schools, and recreational facilities. Fig. 1-2. In the development of these structures, construction technology is helping us satisfy our needs for shelter, education, employment, and recreation, as well as many other needs. The construction of power plants and sewer and water systems meets our utility needs. The construction of broadcasting towers for radio and television stations makes communication within the community and with the rest of the world more convenient.

Agricultural Needs

Construction also provides many agricultural aids. We build wells and irrigation systems to

Fig. 1-1. The construction of a modern skyscraper draws on the full range of available tools in construction technology.

Fig. 1-2. Construction contributes to community development by providing homes and facilities for community services.

irrigate otherwise nonproductive land. Fig. 1-3. For larger applications, we build dams to create reservoirs. Then we can construct irrigation canals from a river or reservoir to supply water to farms in the area. This method is also used to provide water supplies for whole communities.

Transportation Needs

Our transportation needs are also met by construction. Highways, bridges, and airports are built to help people travel from place to place. However, construction does more than just help people travel from one place to another. For example, the construction of railroads and seaports allows us to move large amounts of freight by land or by water.

The construction of pipelines enables us to move resources such as water, natural gas, and oil over long distances economically. The most outstanding example of a pipeline is the Trans-Alaska, or Alyeska, pipeline. Fig. 1-4. This 800-mile (1290-km) pipeline allows producers to transport oil from oil fields in Prudhoe Bay

Fig. 1-4. The Alyeska pipeline was constructed to transport oil some 800 miles (1290 km) across the frigid Alaskan terrain.

in northern Alaska to the Port of Valdez in southern Alaska. Before the pipeline was built, oil companies could not tap the vast amounts of oil in Prudhoe Bay simply because they had no way to transport the oil to other parts of the country.

Fig. 1-3. Irrigation canals can be constructed to supply water to land that otherwise could not be used for farming.

Did You Know?

An aqueduct is a structure for carrying water from one place to another. Though the water usually flowed in open channels in an aqueduct, pipe was sometimes used. In a sense, then, an aqueduct was similar to a pipeline. Ancient Rome was served by several aqueducts. It has been estimated that, in the first century A.D., these aqueducts supplied Rome with about 38,000,000 gallons of water — or about 38 gallons per person. Three of the ancient Roman aqueducts have been repaired for use today.

Fig. 1-5. The construction of sewage treatment plants allows us to remove or neutralize harmful wastes from liquid sewage before the water is returned to lakes and streams.

Sanitation Needs

Even our sanitation needs can be met by construction. The building of sewage treatment plants helps protect our environment by neutralizing or removing harmful wastes from the water we use. The water is then returned to its source. From its source, the water can be reused to irrigate land or to supply water to nearby communities. Fig. 1-5.

Did You Know?

Ancient Rome had a sewage system. Rainwater was drained into the Tiber river by drains built into the surface of the earth. However, by the sixth century B.C., the water was carried to the Tiber through enclosed drains. As well as carrying off rainwater, these sewers also carried off the water that had been used in the public baths.

National Development

The benefits of construction are not limited to individual communities. Construction can also affect the development of a country. For example, the building of the transcontinental railroad, which was finished in 1869, encouraged pioneers to settle the far reaches of the western United States. Fig. 1-6. The railroad made transportation reasonably safe and convenient. After the railroad was completed, people and supplies could be carried from Omaha, Nebraska, to Sacramento, California, in five to six days. Before the construction of the railroad, this same trip took as long as three to four months by wagon. The railroad also improved long-distance communication between people in the East and the West. A letter transported by train could be carried much more safely and quickly than one transported by wagon.

The construction of highway and road systems also has made transportation more convenient. Our highways and roads make it possible to travel by automobile to any of the thousands of towns and cities in our country.

Fig. 1-6. The construction of the transcontinental railroad opened up vast areas of the western United States for settlement.

In addition, these highways and roads comprise a valuable network of pathways that are indispensable in transporting food, materials, and manufactured products by truck. As a result, communities can be established almost anywhere that roads can be built. People can now live farther from their jobs and still be within commuting distance. Fig. 1-7.

Fig. 1-7. The construction of highways and roads has allowed for the growth of cities and towns by making transportation more convenient.

For Discussion

1. In what ways has the interstate highway system changed the way you live? Are you now able to buy food that might have been unavailable before? Are there any other changes?
2. How would the life of a pioneer in the West have been improved through a nearby source of transportation?

FROM CAMELS TO CRANES

The construction technology of today is based on the discoveries made by many people over a period of thousands of years. The first construction tools were handmade, hand-held tools made of rocks or animal bones. Later, people found ways to refine the tools so that they could work better and faster. They also domesticated animals—such as camels, horses, and even water buffalo and elephants—to use as "beasts of burden." For a long time, animals such as these provided the greatest force available for moving heavy building materials around construction sites.

Today cranes, dump trucks, bulldozers, and many other types of equipment help construction workers move and lift large amounts of heavy materials. Fig. 1-8. The conversion from beasts of burden such as camels to sophisticated machines such as cranes did not occur overnight, however. The development of new machines and processes to help us do work is actually the result of many significant advances that took place over a period of many years.

Fig. 1-8. Cranes are now used to lift and move heavy materials.

For Discussion

The text mentions that certain domesticated animals were used to move building materials at construction sites. What limits would the use of such animals place on construction activities and building design?

HIGHLIGHTS OF THE CONSTRUCTION PROCESS

The term **construction process** refers to everything that happens from the decision to build a structure to the owner's acceptance of the completed structure. The construction process involves the owner of the structure and the construction company that builds the structure. Sometimes various other parties—such as attorneys, realtors, and financing companies—also participate at some stage of the process. In this section you will become familiar with the basic steps in planning and building a structure.

Fig. 1-9. This gas station is located near an interstate highway interchange. Why is this a good place to construct a gas station?

Planning a Project

Every construction project begins with several decisions. The first decision is whether or not to build a structure. Once this decision has been made, other decisions are needed. Where should the structure be built? When? How much will it cost? Who will build it? All of these decisions and many more must be made.

Selecting a Site

Choosing the right site or location for a structure is important. For example, fast food restaurants should be built on sites that are easy for people to get to. Fig. 1-9. Factories should be built near good transportation routes. The best location for a school is in the community it will serve.

After a site has been selected, the procedure of acquiring the site begins. Usually a realtor handles the negotiations. When the buyer and seller agree, a contract is signed.

Financing

Financing is the term used to describe the process of obtaining the money used to pay for a project. Construction projects cost a lot of money. A public project, such as a school building, is funded through taxes and bonds. A private project, such as a house, generally is financed by one or two people. Usually, people have to borrow the money to pay for private construction projects.

Did You Know?

Every construction project must be paid for. In the nineteenth century, many American railroads received land from the federal government. This land was along the path of the railroad. Because the railroads then sold most of this land to settlers, they obtained money with which to help finance the building of the railroad. In all, this grant of federal land totaled 131,000,000 acres.

Preparing the Plans

At this time many decisions are made about how the structure should look and how it will function. These decisions are made by architects or engineers and are approved by the owner. The final decisions are written into a set of project plans that identify all the details of the project. Fig. 1-10. The project plans give all the information necessary to build the structure as it was designed and engineered. The plans show the construction workers how to build the structure and what materials to use.

Building the Project

When the final plans are ready, the owner must determine who will do the construction work. The owner hires a construction company, or **general contractor**. The general contractor is in charge of the construction work. When the owner and the contractor agree on the cost of the construction, they sign a contract and construction can begin.

Organizing the Work

After the contract is signed, the job of organizing begins. The organization of a construction project is accomplished in part by scheduling.

Scheduling involves estimating the amount of time it will take to do each part of the job. The contractor prepares a schedule for the work that identifies who will do what job and in what order. Fig. 1-11.

Next, the job site must be organized. This includes making sure that workers have access to the site. Portable offices and restrooms, as well as temporary utilities such as water and electricity, are needed.

Completing the Project

As the job progresses, managers must make sure that the construction tasks are being done properly and on time. The contractor keeps charts and records of the progress. The object is to get the job done on time and to stay within the established budget.

Fig. 1-11. This construction manager uses a bar chart to help schedule construction tasks in the right order.

Fig. 1-10. The project plans contain all the information necessary to build a structure.

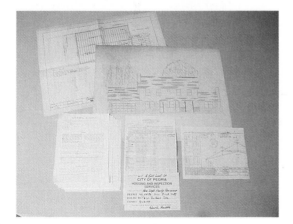

Completing the project includes finishing the structure itself as well as completing all of the exterior work, such as paving and landscaping. It also includes cleaning up the construction site after the building is finished.

When all of this work has been done, the project is ready for the final inspection. Inspectors check to see that the job has been done properly and according to the contract. Fig. 1-12.

During the inspection, the inspectors make up a list of things to be corrected. The contractor has to make the necessary corrections by a certain date. Once the corrections are completed, another inspection is made.

When everyone is satisfied that the contract requirements have been met, the owner accepts the completed project. This is called the **transfer of ownership**. It includes a formal notice of completion, which legally establishes that the job is done. At this time the owner makes the final payment to the contractor. Now the owner takes on the responsibility for the maintenance and repair of the structure and grounds.

Fig. 1-12. These people are performing the final inspection for this building. They will note anything that has not been done correctly.

For Discussion

Discuss the qualities a person should look for in a general contractor.

HEALTH & SAFETY

Safety on a construction site is important. You may have noticed that fences are sometimes placed around large construction sites. As construction technology has developed, the number of regulations covering the safety of construction workers has increased. Workers today are provided with more protection on the job than were construction workers one hundred years ago. Much of this increased protection has resulted from state and federal laws. Also, consumer groups are watchful for unsafe job practices.

THE PRICE OF PROGRESS

Much progress has been made in the construction industry since our ancestors used stone or bone tools to construct their dwellings. In general, most people agree that the progress we have made in construction technology has been useful and good. However, construction is not without some drawbacks. For example, every time we clear a tract of land for a new construction project, we alter, or change, a part of our natural environment. Figs. 1-13 and 1-14.

In many cases the construction project destroys the natural habitat of a variety of plants and animals. As a result of the widespread building to accommodate human growth, many plants and animals have become extinct. Therefore, the possible effects of a proposed construction project should be weighed carefully before the decision is made to begin building. By making responsible decisions, we can build structures that will do as little harm as possible to the environment and still provide for our needs.

Fig. 1-13. Even though most construction projects are beneficial to people, we must consider whether their benefits outweigh any possible negative effects on our environment.

Fig. 1-14. The water behind this dam covers land that was once part of several small family farms.

For Discussion

Are you aware of any construction projects in your neighborhood that have been built on land that was covered with trees or bushes? Did the builders make any attempt to keep some of the trees or bushes? Did they plant any new trees or bushes? In your opinion, were any trees or bushes removed unnecessarily?

Construction Facts

A GOOD ARCH TAKES TIME

A monumental symbol of the twin spirits of freedom and exploration, the Gateway Arch in St. Louis is the tallest freestanding arch in the world. The stainless steel arch stands 630 feet (190 m) high—as tall as a 63-story building.

In the late 1940s, the people of St. Louis saw the need for a memorial to commemorate the Louisiana Purchase of 1803 and the westward expansion that resulted. A national competition was held in 1947 to select a design for the monument. A deceptively simple arch, the entry of architect Eero Saarinen, was the winner of the competition.

Saarinen's original design presented severe engineering problems. For example, his original design called for a nonstructural stainless steel covering over structural steel members. With this design, the arch could not withstand the forces to which it would be subjected. Even a light wind would twist the arch and break it off at its base.

It was not until 1961 that construction began on a perfected arch. However, many new problems were yet to be encountered. For example, special crawler derricks had to be used to carry the upper sections of the arch into place. Also, at the 530-foot level, a stabilizing strut had to be placed between the two legs of the arch.

In 1965, seventeen years after its conception, the arch was completed. Large jacks were used to push the legs of the arch apart so that the final section, the keystone section, could be inserted.

The finished arch has withstood the test of time. For more than 20 years it has towered over St. Louis as a huge symbol of the role of that city as the "Gateway to the West." This beautiful and masterfully engineered structure is considered by many to be an architectural triumph.

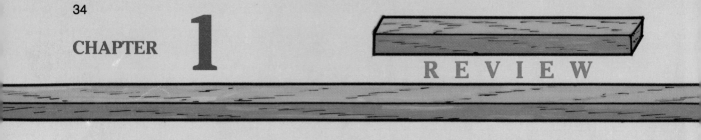

CHAPTER **1**

R E V I E W

Chapter Summary

The construction industry today reflects developments in technology occurring over thousands of years. Today's construction technology helps to meet our agricultural, transportation, and sanitation needs. Construction, especially as it relates to transportation, has also been important in national development. Construction processes must be carefully planned. A site must be selected. Financing must be obtained. Plans must be prepared and the work must be carefully organized by the contractor in charge of the project. In all building projects, it is essential that as little harm as possible be done to the environment.

Test Your Knowledge

1. Name at least five types of structures that can be built to aid in the development of a community.
2. What types of structures can be built to provide agriculture aid in areas where irrigation is necessary?
3. Name four types of structures that can be built to help meet our transportation needs.
4. How can our sanitation needs be met by construction?
5. Explain how the construction of the transcontinental railroad affected the development of the western United States.
6. Name three types of equipment that construction workers use today to help lift and move heavy materials.
7. Who are the two principal participants in any construction project? Who else may be involved?
8. What is the function of the project plans for a construction project?
9. Name three tasks that are included in organizing a job site.
10. Why must we consider carefully all of the effects of construction before we begin to build?

Activities

1. Choose a construction project that was recently completed in your community. Make a presentation to your class to explain the effect of that construction project on your community.

2. Do research on a major construction project such as the Erie Canal, the Alyeska pipeline, or Hoover Dam. Write a one-page report on the effects of the construction project on the environment and on society.

3. Use old newspapers and magazines to create a construction-related collage on a piece of cardboard. Design your collage so that it illustrates one of the following themes:
   - Community development
   - National development
   - Progress in construction
   - Process of construction
   - Environmental impact of construction

2 CONSTRUCTION AND SOCIETY

Terms to Know

built environment
construction
ecology

environment
environmental
    impact study

Objectives

When you have completed reading this chapter, you should be able to do the following:

- Define *built environment*.
- Explain how construction contributes to the economy.
- Identify two major concerns people have about construction.
- Identify several factors that affect the cost of a construction project.

Construction, or the building of structures, plays an important part in our lives. Through construction we are able to meet our basic need for shelter. Homes, apartment buildings, and hotels are common examples of buildings that we construct to provide us with shelter. Through construction we also build office buildings, factories, and other buildings that, in addition to providing shelter, provide us with places to work. Construction is also the process by which we build roads, highways, bridges, and tunnels that we use in transporting people and products from place to place. As you can see, construction plays an important role in our society. Construction is a primary activity by which we shape our environment to better suit our needs.

HEALTH & SAFETY

Did you know that a building can be "sick"? Some new and remodeled office buildings contain materials that give off harmful fumes. In some cases, this problem is worsened because the building is closed. Because the windows do not open, the toxic fumes build up within the building. It is estimated that as many as 10 percent of the workforce are sensitive to such fumes. In some cases, the solution is clear. The material giving off the harmful fumes (carpet, for example) can be replaced. In other cases, the solution is not as simple.

THE BUILT ENVIRONMENT

Environment simply means "surroundings." Most often when we think of the word *environment*, we think of trees, lakes, and mountains. However, structures such as buildings, bridges, and highways are part of the environment, too. These structures, along with other parts of the environment that people have shaped or altered can be referred to as the **built environment**. Fig. 2-1. Our built environment is ever changing to meet current needs. Old buildings are often renovated (renewed). Other buildings are torn down and new buildings take their places.

Fig. 2-1. The fountain, sidewalks, and other structures in this park were built by people. These structures are part of our built environment.

Personal Benefits

We all have the common needs for food, clothing, and shelter. The need for shelter is satisfied directly by construction. For example, we live in houses and apartments that shelter us from the weather. Fig. 2-2. Some of these houses and apartments are fancier and more expensive than others. Each has been constructed with the basic elements of a foundation, walls, and a roof.

The construction industry indirectly satisfies other needs, such as needs for food and clothing. For example, stores, factories, and warehouses are constructed to provide manufacturing space and places to store and sell clothing. Food processing plants are constructed that make it possible to process large quantities of food in short periods of time. This helps us keep up with the ever-increasing demand for food. Once the food is processed, it is stored and sold in buildings that have been constructed for that purpose. Fig. 2-3.

Fig. 2-3. Grocery stores are specialized buildings designed and constructed to store, display, and sell food.

Fig. 2-2. These apartment buildings were constructed to shelter several hundred families.

Food and clothing do not depend on construction directly. They would still be available if stores and factories could not be built. However, the existence of these buildings makes food and clothing more accessible and provides us with better-quality products.

Social Benefits

In addition to meeting personal needs, construction affects our social lives. People are social beings. We like to interact with one another. Living in a community makes such interactions possible. Playgrounds, public swimming pools, and other recreational facilities are just a few products of construction that help satisfy our need to socialize. Fig. 2-4.

Construction also helps us share our experiences and culture. The construction of large facilities allows people to gather at cultural functions such as concerts, ballets, and plays.

Fig. 2-4. Recreational facilities such as this ski lodge are constructed to help us meet our social needs and wants.

Fig. 2-5. A museum offers a place for us to learn about the past.

Museums, libraries, and art galleries help people learn more about their society. Fig. 2-5. Community centers and auditoriums provide places for people to gather to decide important issues or to share interests.

Communication

Because of fast-changing technology, communication systems are changing and improving rapidly. We can communicate over long distances conveniently and with great speed. Our communication systems are so sophisticated that we can communicate with people across the world without leaving our homes. Television broadens our world by bringing news and informative programs about the world directly into our living rooms. Fig. 2-6.

Our communication systems require the construction of transmission towers, relay stations, and radio and television studios. Many people are involved in designing and constructing the

Fig. 2-6. Construction plays an important role in many types of communication. This television broadcasting studio required the construction of many unusual structures.

different components of communication systems. Architects, engineers, contractors, and other workers contribute to the construction of these systems.

Transportation

Transportation systems seem to have made the world smaller and each person's horizons wider. Because of our excellent transportation systems, we can live and travel practically anywhere. Railways, highways, bridges, and airports are construction projects that make travel fast and convenient.

When you think of transportation systems, you probably think first about the vehicles that carry us from place to place. However, in many cases the pathways are as important as the vehicles in providing safe and efficient transportation. For example, trains could not operate without a system of tracks. Automobiles would travel more slowly and be less convenient to use if we did not have an effective system of roads and highways. This system of roads and highways represents major construction efforts. In the United States there are over 4 million miles (6.4 million km) of roads and 300 thousand miles (480 thousand km) of railroad tracks. Fig. 2-7.

Fig. 2-7. The construction of an effective road and highway system has made travel by automobile more efficient.

Did You Know?

The wheels of modern trains are, of course, guided by rails. In the Bronze Age, people in the eastern Mediterranean constructed ruts in roads. These ruts were 3 to 6 inches (76 to 152 mm) deep, about 8 inches (203 mm) wide, and about 50 inches (1.27 m) apart. The function of the ruts was the same as the function of the rails in a railroad—to guide the wheels. The ruts were used to guide the wheels of carts and wagons. Of course, this made the transportation of goods easier.

For Discussion

1. In what ways has the design of buildings been influenced by the fact that human beings are social creatures?
2. Why are rivers in the United States no longer greatly used for passenger travel?

CONSTRUCTION AND THE ECONOMY

The construction industry affects our nation's economy by employing millions of people. Within the construction industry are many different types of jobs. The obvious jobs of carpenter, plumber, and electrician are not the only jobs offered by the construction industry. Construction workers also include architects, engineers, accountants, and managers. Each job provides workers with income that allows them to buy the things they need and want. This process of earning and spending money contributes to the health of our economy.

Construction and community growth are closely related. As a community grows, the number of available jobs also grows. Constructing a new factory or business attracts workers to a community. Fig. 2-8. When new workers and their families move to a community, they purchase houses and use the community's goods and services. They generate business that improves the economy. As a community grows, more highways and utilities are needed. Thus there is a need for more construction.

Fig. 2-8. When this new factory is completed, additional jobs will be available for people living in nearby communities.

Fig. 2-9. The damming of water by the Aswan High Dam in Egypt has had some harmful effects on Egyptian agriculture.

ECOLOGY AND THE ENVIRONMENT

Ecology is the study of the way plants and animals exist together. It also studies the relationship of plants and animals to their environment. The changes we make that affect our natural environment can be critical. If we make a poor decision, we may damage our environment beyond repair. For example, in the 1960s, the Egyptians built the Aswan High Dam on the Nile River. Fig. 2-9. The Aswan dam is 364 feet

(111 m) high. According to those who favored the dam, it would control seasonal flooding of farmland and residential areas. It would provide electricity for half the population of Egypt. It would also create approximately 1 million acres (404,680 ha) of new farmland through irrigation. When the dam was built, these benefits did occur. However, many unforeseen environmental effects also occurred.

Before the dam was built, the seasonal floods brought new nutrients to the farmland every year. Now that the floods have been controlled, the land is beginning to lose its fertility. The salt level in the land has risen by 10 to 15 percent, making it unsuitable for growing some kinds of crops. Another effect of the dam is coastal erosion, which has required the construction of costly dikes to keep the land from washing away.

The Aswan High Dam has also affected Egyptian culture. The dam has caused the water table to rise, threatening to cover monuments such as the Temple of Karnak. Fig. 2-10. Many of the Nubian people whose ancestors lived along the banks of the Nile, have been displaced. To support themselves, these people will have to learn to farm the irrigated land.

Fig. 2-10. The Temple of Karnak is one example of ancient Egyptian architecture that is threatened by the rising water table in Egypt.

Did You Know?

The soil in the Nile delta is some of the richest soil on earth. For centuries it has provided Egyptian farmers with land on which they could grow crops. The Aswan High Dam, completed in 1971, slowed the flow of the Nile through the delta. This allowed the waters of the Mediterranean Sea, which is saltwater, to flow farther up the Nile. As a result, the delta land covered by this seawater absorbed some of its salt. This salt in the soil lessens the usefulness of the land for farming.

The Aswan High Dam is only one example of how an uninformed decision can have unpredicted or harmful effects on our environment. In the United States, research committees study the effects of proposed major construction to try to identify in advance the harm it may cause. We also have a law, the Endangered Species Act, that helps protect animals and plants in danger of extinction because of human activities.

Political pressures, however, often bring about exemptions from and exceptions to the rules.

In Tennessee, for example, the construction of the Tellico Dam caused a great deal of controversy. Research committees found the dam to be economically unfeasible. Furthermore, 5,600 acres of agricultural land would be lost if the dam were built, as well as 280 archeological sites listed in the National Register of Historic Places. Also, scientists discovered that a species of minnow, the snail darter, would be threatened with extinction. The dam would remove the snail darter's only existing natural breeding ground. Fig. 2-11.

The dam would supply $2.7 million in electricity annually. The reservoir created would provide a recreational area. Because of these factors, political pressures to build the dam were strong. The protests of the environmentalists were overruled, and the dam's builders were exempted from complying with the Endangered Species Act. The agricultural land and the historic sites were lost. Fortunately, scientists were able to transplant 2,000 snail darters to another river in Tennessee, where they seem to be doing fine. If the transplant had not been successful, a whole species of life would have been destroyed.

As we plan new structures to meet our needs, we must give thought to both the built and the natural environment. The effects of a proposed construction project on the natural environment need to be evaluated. We must consider the preservation and conservation of open space, water resources, air quality, and animal and plant life.

A study that evaluates the effects of a project on the environment is called an **environmental impact study**. The environmental impact study is meant to bring to light the effect of a construction project on the environment. The federal government requires environmental impact studies on almost all construction projects that are built with federal money. Fig. 2-12.

Fig. 2-11. The snail darter is an example of a fish threatened by extinction due to our construction activities.

Fig. 2-12. The area shown here has been contaminated by industrial contaminants. These ecologists are measuring its effects. To prevent damage to the environment, an environmental impact study is required for any new construction project that uses federal money.

In many cases construction can improve the natural condition of the environment. For example, if mud slides often *erode* (wear away) a hilly area, retaining walls can be built to hold back the soil. Other construction projects, such as a water treatment plant, can improve our natural resources so that a community can use them. Fig. 2-13. Such a construction project, if well planned, can complement the natural environment. Each situation has to be evaluated on its own merit.

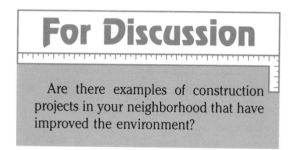

For Discussion

Are there examples of construction projects in your neighborhood that have improved the environment?

COSTS OF CONSTRUCTION

Most construction projects are very expensive. The high cost of construction is having a profound impact on our society. To understand the impact it is necessary to understand what makes up construction costs.

One large construction expense is the land on which a structure is built. The cost of the site for the average house is about 20 percent of the house's total cost. Also, construction projects must be built with sturdy, durable materials. The costs of structural materials—lumber, steel, and concrete—constitute a major building expense. Fig. 2-14.

Many other factors also contribute to the overall cost of a construction project. Construction workers must earn a living wage. *Overhead*, or the cost of running the company, and the company's profit must also be included in the cost.

Fig. 2-13. The construction of a water treatment plant provides a place where water can be made safe for human consumption.

Fig. 2-14. Traditional building materials such as wood are becoming more expensive. Their costs contribute to the overall cost of construction, which is rising steadily.

Because of the high cost of construction, many people who probably would have purchased houses ten years ago now live in apartments or condominiums. In many cases, condominium and apartment dwellers share common walls with neighbors on both sides, upstairs, and downstairs. These conditions require that the people be considerate so as not to disturb one another's privacy. Even a loud alarm clock can sometimes be heard in the room next door. Insulation in the walls does not always eliminate all of the noise.

Although apartments and condominiums provide less privacy, they often include recreation facilities. For example, many apartments and condominiums have their own swimming pools and tennis courts. Fig. 2-15. Also, apartment and condominium dwellers rarely are responsible for yard work and for maintenance and repair of the building.

As construction costs continue to rise, more and more people will be affected by these and similar changes. Also, pressure will increase to keep costs low. New, less expensive materials undoubtedly will be developed to take the place of more expensive ones. These new materials may even change the appearance of the buildings we construct in the future.

For Discussion

In your neighborhood, there may be buildings of many different styles. The buildings may be constructed using a variety of materials. Discuss some of the reasons for the use of various styles. Discuss also the reasons why some building materials may have been chosen over others.

Fig. 2-15. The individual units in this condominium building are built around a pool. Those who live in the condominium have access to the pool.

Construction Facts

Years ago, every town had a town square or village green. This served as a central meeting place for townspeople and a place for special events. The town square served a real need in the community. Now we have shopping malls. Look around you next time you are in a shopping mall. We shop there, certainly, but we also meet and socialize there, just as townspeople in the past socialized in the town square.

The shopping mall as we know it today came into being in 1956 in Edina, Minnesota. Southdale Mall, designed by Victor Gruen, was the first fully enclosed shopping mall.

Gruen was an Austrian architect working in America. He believed that people spent too much time isolated, driving here and there. He thought that a community needed a place for people to be together face to face. Also, he saw a way to use land more efficiently by grouping stores together instead of placing them throughout the city. Gruen's idea for enclosing the mall solved the weather problems for shoppers, too.

The success of the Southdale Mall proved that Gruen's idea worked. In fact, his idea worked so well that Southdale Mall became the model for new malls all across the nation. Gruen knew that the mall could meet a social need by providing a center for human activity that was exciting, comfortable, and fun. There are now hundreds of shopping malls across the country. The mall shown here is in Texas.

CHAPTER **2**

REVIEW

Chapter Summary

Construction plays an important part in our lives. Through construction, we alter our environment by creating the built environment. Construction creates personal benefits and social benefits by improving communication and transportation. It affects the economy by employing millions of people. Construction projects also affect the environment. Before constructing a project, possible harm to the environment should be carefully studied. Construction can be expensive. The cost of construction must include the price of the land and materials, as well as the cost of labor. The increasingly high cost of construction may lead to the development of less expensive building materials.

Test Your Knowledge

1. What is an environment?
2. Name three personal benefits that construction makes possible either directly or indirectly.
3. How does construction contribute to our social lives?
4. Name two ways in which construction affects the economy of a community.
5. Name three beneficial results that building a dam might have.
6. Name three harmful results that building a dam might have.
7. What law in the United States is aimed at preserving plants and animals threatened by extinction?
8. What is the purpose of an environmental impact study?
9. Name three major types of costs that are involved in a construction project.
10. Name one major effect that the high cost of construction has had on our society.

Activities

1. Choose a construction project that has had a profound effect on the environment. The project may be a dam, a nuclear power plant, a canal, or a similar project. Find newspaper and magazine articles that discuss the effects of the project on the environment. Share your findings with your class.

2. Do research to find out more about environmental impact studies. For example, find out what a study tells about a proposed project and how a study affects the final decision to build or not to build a structure. Write a one-page report about your discoveries.

3. Find a place in or around your community where the natural environment could be improved by construction. Describe the improvement you think could be made and how it would affect the natural environment.

4. In 1980 a rare subspecies of butterfly, the Palos Verdes Blue butterfly, was placed on the federal government's list of endangered species. This rare butterfly lived only in a small area of land, overgrown with locoweed, south of Los Angeles on the Palos Verdes peninsula. In 1983 the city of Rancho Palos Verdes destroyed the butterfly's habitat to build an athletic playing field. The tiny creature is now thought to be extinct. Write a one-page report on the Palos Verdes Blue butterfly. Discuss whether or not the playing field should have been built.

CHAPTER

3 CONSTRUCTION SAFETY

Terms to Know

accident   safety factor
first aid   safety rules
labor unions
Occupational Health
  and Safety
  Administration
  (OSHA)

Objectives

When you have completed reading this chapter, you should be able to do the following:

- Identify two major safety concerns in the construction industry.
- Describe the safety measures taken by construction companies to protect workers on the job.
- Understand the importance of safety rules and regulations in your school laboratory.
- Describe first aid techniques to use in case of an accident.

Safety is a major concern in the construction industry. While the potential for accidents exists in any business, the nature of construction increases the possibility of accidents. Therefore, no study of construction would be complete without a review of safety.

The two safety issues that most concern construction workers are safe working environments and building structures that are safe for people to inhabit or use. The primary concern of safety programs in the construction industry is to prevent injury to people and damage to property. In this chapter you will learn how the construction industry endeavors to meet the needs for safety in these areas. You will also learn how you can practice safe construction techniques in your school laboratory.

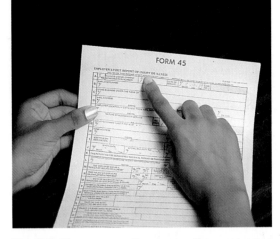

Fig. 3-1. Most construction companies require that an accident report be filled out for any accident, no matter how minor. The accident report helps supervisors correct problems and maintain a safe environment for employees.

ACCIDENT PREVENTION

An **accident** is an unexpected happening that results in injury, loss, or damage. Some accidents are minor, such as getting a splinter in your finger. Other accidents are serious, causing severe property damage, personal injury, or even death. All accidents, both minor and severe, can and should be prevented by taking appropriate safety precautions.

Every accident has a cause, or reason why it happened. Fig. 3-1. An accident may be caused by the carelessness or forgetfulness of a worker. For example, a worker who is hammering nails into a board and is not paying attention to what he or she is doing might hit a finger with the hammer. Accidents may also be caused by faulty equipment. For example, a broken electrical cord on a power tool might cause someone to get an electrical shock. Fig. 3-2. These are only a few examples of how an accident may be caused.

Fig. 3-2. A frayed electrical cord presents two safety hazards: it can give you a severe electrical shock, and it is a potential fire hazard.

Whenever an accident happens, regardless of its cause, it has an effect on someone. A worker may be hurt, or there may be equipment or property damage. Of course, time is lost also, which puts the work behind schedule. As a result, all accidents cost money. The cost of construction rises because of accidents, so that the structure costs more in the long run. Also, insurance rates are higher, which contributes to the overall cost of a project.

For Discussion

Safety rules may be general and refer to work practices as a whole. Safety rules may also be specific and refer to a certain work practice (such as working with electrical wiring). List three general safety rules.

CONSTRUCTION SAFETY

The primary concerns of safety in construction are to prevent injury to people and to prevent damage to property. There are two major ways to assure safety in the construction industry. One is to design safe structures. The other is to make sure that the workers have a safe place to work and know how to work safely. The workers then have the responsibility to follow company safety regulations to prevent accidental injury or damage to property.

Designing Safe Structures

All structures must be designed so that they are safe to inhabit and use. Fig. 3-3. For exam-

Fig. 3-3. This building was carefully planned by engineers so that it will be able to bear the weight of the construction materials as well as that of the people and things it will contain.

ple, buildings must be strong enough to support their own weight as well as the weight of all the people and things inside. Bridges must be wide enough and have strong enough foundations to support the heaviest vehicles they will carry. Roads, water towers, buildings, and every other kind of construction project must also be designed for safety.

Did You Know?

The ancient Romans thought that architecture was the art of building. They thought that architecture was concerned mainly with three ideas. The first of these was the stability, or safety, of the building. The second was the appropriate use of space in the building. The third was the attractiveness of the building.

To make sure that structures are strong enough for their intended uses and will not fail, engineers and architects use a safety factor. A **safety factor** is an extra measure of strength added to the design of a structure. For example, if an elevator must be able to carry ten people, the engineer may actually design the elevator to hold fifteen people. This safety factor helps prevent the elevator from failing due to overload.

Promoting Worker Safety

Most construction companies create and enforce a set of company safety rules. **Safety rules** are regulations aimed at preventing accidents and injuries in the workplace. The company's safety rules are taught to new workers before they begin working at a construction site, or _job site_. Special training sessions also may be held for workers already on the job. These sessions inform the workers about proper safety practices.

The safety rules developed by construction companies are usually just common sense. Most construction companies require workers to wear hard hats. These hats protect the workers' heads from falling objects. Fig. 3-4. Ear protection may be required in some construction areas to protect against harmful noise such as that from a concrete saw or jackhammer. Safety shoes protect workers' feet from the danger of heavy

Fig. 3-4. Most construction sites are "hard hat areas." This means workers must wear hard hats to protect themselves from falling objects.

objects being dropped on them. Safety belts and harnesses with life lines attached should be worn by workers who work at high levels.

Promoting worker safety also includes maintaining a safe job site. The use of equipment guards, caution signs, and safety storage cans are good ways to maintain a safe job site. Fig. 3-5. Maintaining clean work areas that are free of clutter also helps promote worker safety.

Another way in which a construction company can help create a safe working environment for employees is to maintain and repair company tools and equipment. This requires a routine inspection of all tools and equipment. Damaged and worn tools and equipment must be repaired or replaced because they present a health hazard to employees.

In addition to the construction companies, three other groups are involved in promoting worker safety. These groups are the United States government, insurance companies, and labor unions. These groups have different functions, but they share the common goal of preventing accidents on the job site.

Fig. 3-5. The guard on this circular saw covers the teeth on the blade to protect the worker's hands.

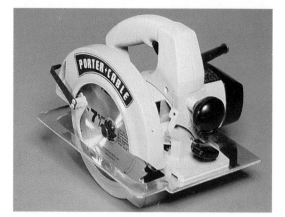

The Government

In 1970, the U.S. Congress created a regulatory agency known as the **Occupational Health and Safety Administration (OSHA)**. OSHA sets standards that regulate safety at construction sites. OSHA inspectors check on companies to make sure they are following these safety regulations. OSHA also inspects and approves safety products such as safety glasses and storage cans. If an OSHA inspector finds unsafe conditions at a job site, he or she informs the owner of the company. The inspector is authorized to shut down the construction job until the conditions are corrected.

Did You Know?

Lack of safety was a problem with early elevators. Prior to 1850, elevators were used mainly for lifting freight. These elevators were lifted by ropes, which sometimes broke. In 1853, Elisha Otis introduced an elevator that used a safety device to prevent it from falling. This safety device used a type of clamp that gripped the elevator guide rails. This clamp was activated when tension was lost on the hoisting rope. It was this basic development that opened the way to the development of the passenger elevator.

Insurance Companies

Insurance companies also are concerned about safety in construction. If a worker is injured on the job, an insurance company may have to pay his or her medical expenses. Insurance companies keep accurate records of accidents that happen at construction sites. The insurance companies use these records to set *insurance rates.* These determine how much the company's insurance will cost. By working to prevent accidents, a construction company can keep its insurance rates low.

Labor Unions

Labor unions are worker-controlled organizations that are formed to present the demands of the workers to the management of construction and other types of companies. Workers form unions to increase their bargaining power. As members of a union, they are better able to negotiate for higher wages, shorter working hours, and improved working conditions.

Labor unions also negotiate for safer working conditions. For example, union workers in a job site may notice several safety hazards. The union may ask the construction company to eliminate these hazards. If the company managers take no action and the safety hazards are severe, the union workers may go on strike. To prevent workers from striking, the company may have to meet the workers' demands by increasing its safety measures.

Did You Know?

One of the first trade unions was formed in Tolpuddle, a small village in England. In 1833, six English farm laborers from that village formed a trade union. The purpose of their small trade union was to prevent their wages from being reduced. The laborers were arrested, and sentenced to banishment to a penal colony in Australia. The charge was "administering unlawful oaths." There was an enormous public outcry against this sentence. As a result, the men were set free in 1836. All but one of them returned to England.

For Discussion

Can you name some of the features in the design of a school building that help make it a safe structure?

DEVELOPING A SAFETY PROGRAM

The best plan for accident prevention is to have a company safety program. The goal of a safety program is to achieve longer and longer periods of time without injury. One of the most important elements of a safety program is to develop good attitudes of safety among the workers. Then workers must be encouraged to follow general and specific safety rules concerning their personal safety while using tools and equipment. Fig. 3-6.

Fig. 3-6. This sign reminds the workers to work safely.

A company safety program also includes concern for the safety of the general public. Fig. 3-7. Sometimes security guards or security fences with locked gates can help ensure safety. Safety barriers, fences, and temporary protected walkways can be used to protect passersby from any danger at the construction site. The public should be allowed to view the project, but from a safe vantage point. This will ensure safety for spectators and good public relations for the company.

In the following paragraphs, you will learn more about the elements of a good safety program. You will find parallels between safety on a construction site and safety in your school laboratory. Read this information carefully before you begin any construction work in your laboratory.

Safety Attitudes

A good positive attitude toward a safety program or safety on the job is essential to the success of the program. If safety procedures and safeguards are not important to the worker, then the worker will disregard the rules and will not follow safe working procedures.

In most companies, the development of safe attitudes begins in employee training sessions. These sessions encourage safety attitudes as well as give instruction on safety practices. Much depends on a worker's attitude. You can give a worker safety instruction, but you cannot make the worker apply what he or she has learned. The worker must have the right attitude.

A worker who observes good personal safety practices is to be commended. Good personal safety practices are not enough, however. A worker with a good safety attitude must show concern for the safety of others. A safe worker is part of a safe team. Teamwork is an important element in safety. When team members are working safely together as a whole, then each member feels a responsibility to the team and wants to work safely for the good of the team. Fig. 3-8.

Fig. 3-7. This covered walkway protects pedestrians as they walk near the construction site.

Fig. 3-8. Because construction workers work together as a team, each team member must think about everyone's safety.

HEALTH & SAFETY

In the nineteenth century, a worker who was injured on the job could do little about it. Generally, the employer was not responsible. However, in the late nineteenth century, there was an increased awareness of the need for safety in the workplace. At the same time, the employer began to take more responsibility for on-the-job injuries to employees. As a result of this, the employer sought to make the workplace safer for the employee. This resulted in a great decline in on-the-job injuries. As one example, accident rates for machine hazards are now only one-half of what they were forty years ago. This lowered accident rate is due mainly to the installation of protective devices on machines.

General Safety Rules

Most of the general rules for safety are just common sense rules. These rules are the same for your school laboratory as they are for employees on a construction site. Violating these safety rules will endanger you and everyone else in the laboratory area. The following is a list of general safety rules you should know before beginning work in your construction laboratory. Following these rules may help you avoid serious injury.

1. Make sure an instructor is present in the laboratory area before you begin your work.
2. If you do not know how to operate a tool or machine, ask your instructor to help you.
3. Concentrate on your work at all times. Running in the laboratory area is not permitted. Before you begin any operation, carefully think through every step that you will take. Identify any possible threats to your safety,

and take immediate action to remove them. Do not take chances. To do an operation safely often takes more time. However, it is better to spend time on safety than to suffer a serious injury. Remember, you can prevent most accidents by thinking *safety first!*

4. Keep your work area and all passageways clean and free of rubbish.
5. Keep your equipment in proper working order.
6. Stay alert when working with machinery.
7. When you are working with materials that may chip and fly into your eyes, wear safety glasses or goggles to protect your eyes. Fig. 3-9.
8. Wear hearing protectors or ear plugs in noisy work areas. These devices will help prevent your hearing from being damaged by excessive noise. Once damaged, your hearing cannot be restored.
9. Wear a respiratory mask when you are working in an area filled with dust or fumes. Fig. 3-10.

Fig. 3-9. Safety goggles provide eye protection. They help keep chips from flying into a worker's eyes.

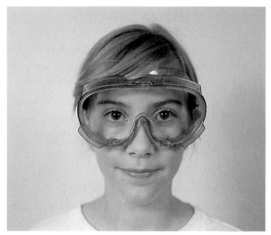

Fig. 3-10. Respiratory masks help protect workers from inhaling harmful particles.

10. Dress properly before you begin work. Remove jewelry such as necklaces, bracelets, watches, and rings. It is important to wear shoes at all times. Open-toed shoes such as sandals do not protect your feet from flying sparks or from sharp objects on the floor. Long hair should be tied back or placed securely under a cap. Hair can get caught in tools and machines and cause serious injuries.
11. Keep your mouth free of food and other objects. Chewing gum, toothpicks, and fasteners such as nails are hazardous to have in your mouth when working. An accidental slip or fall could cause you to choke on such items.
12. Use caution when lifting and moving objects. You can severely injure your back by lifting materials that are too heavy. Always ask for help when lifting heavy materials or carrying long objects. Lift using the muscles in your legs, not in your back.
13. Obey all safety signs in the laboratory area.
14. Be courteous to fellow students in the laboratory area. Arguing wastes time and energy, and may distract other workers in the laboratory area. Be willing to help others.

15. Report all accidents to your instructor immediately. An accident report form should be filled out after an accident has occurred.
16. Know who the trained first-aid person is and where the first aid kit is.

Tool and Equipment Safety

Accidents involving tools and equipment generally occur because the operator has violated a safety rule. Many of these accidents can be avoided by observing the following safety rules:

1. Use tools and equipment only after you have received proper instruction in their use.
2. Use each tool and piece of equipment only to do the job for which it was intended. Do not attempt to modify tools or equipment.
3. Handle all tools and equipment with care. It is not safe or smart to throw tools down on work tables or to abuse equipment. Such actions can cause injuries or damage tools and equipment.
4. Carry tools in your hands, not in your pockets. Many tools have sharp edges and points that might cut you, especially if you were to slip or fall. It is also unsafe to carry too many tools at once.
5. Always check the electrical cords and plugs on portable power tools before using them. Cords should be in good condition with no frayed edges. Three-prong plugs should have all prongs intact. If the grounding prong is missing, the tool could give you a severe electrical shock. Also keep the cord well away from any cutting blades. Fig. 3-11.
6. Unplug portable power tools such as drills and sanders when they are not in use. Before plugging in a power tool, make sure all operating switches are off to keep from starting the tool accidentally.
7. When you operate portable power tools, be aware of your surroundings. Do not stand in water or work near live electrical wires.

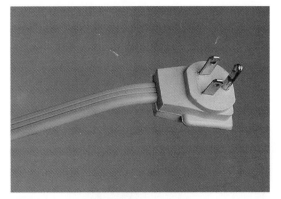

Fig. 3-11. This plug must be plugged into a grounded outlet to protect the worker from electrical shock.

8. Keep all guards in place, even when equipment is not in use.
9. Never touch moving parts on equipment. When you adjust equipment, make sure all switches are off and all moving parts have stopped.
10. Make sure a machine has completely stopped running before you leave the work area.
11. Report all damaged tools and equipment to your instructor to avoid accidents. If a tool or machine malfunctions, turn it off immediately.

Fire Safety

Fire prevention is an important part of any safety program. Since you may be working with potential fire hazards, it is important for you to learn some basic rules of fire safety. The safest procedure is to turn in a fire alarm regardless of the size of the fire and then promptly use a fire extinguisher to put the fire out. Emergency fire fighting equipment should be available in convenient locations. You should know where it is located.

The laboratory should be free of accumulated rubbish. Passageways should be clear of obstacles so that the exit is easily accessible if a fire should occur. All extinguishing equipment should be in its proper location, clearly marked, and in working order.

The following rules will help you prevent fires from starting. They will also give you a clear idea of what to do in case a fire breaks out:

1. Dirty rags soaked with oil, grease, or paint are a serious fire hazard. Place dirty rags in a proper safety can after use.
2. Flammable liquids such as paint and varnish should be stored in a well-ventilated place away from all furnaces and other heat sources. Fig. 3-12.
3. Make sure all electrical cords are in good condition. Frayed cords are a fire hazard.
4. Become aware of the location of fire extinguishers so that if a fire breaks out, you will not waste time searching for one.

5. Learn the proper way to use a fire extinguisher. Fig. 3-13.
6. Not all fire extinguishers can be used to put out all types of fires. For example, a water-type extinguisher should *never* be sprayed on an electrical fire. Figure 3-14 shows the four main classes of fire extinguishers—A, B, C, and multipurpose.
7. If a fire starts, quickly activate the nearest fire alarm to alert others of the danger. Then, if the fire is small and not spreading quickly, you may be able to control it with a fire extinguisher. If the fire is spreading quickly or cannot be extinguished, leave the building immediately.
8. If your clothing catches on fire, *do not run!* Drop quickly to the ground and roll from side to side to extinguish the flames.

These are basic fire safety rules that apply to all construction sites. Like other accidents, most accidental fires can be prevented by careful planning and foresight.

Always remember to *think safety first!*

Fig. 3-12. Store flammable materials such as paint and varnish in a dry, well-ventilated area away from heat.

Fig. 3-13. The correct way to use a fire extinguisher.

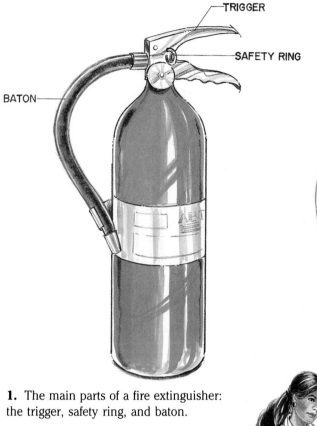

2. Holding the fire extinguisher firmly, pull out the safety ring.

1. The main parts of a fire extinguisher: the trigger, safety ring, and baton.

3. Pick up the extinguisher with one hand on the trigger. Use the other hand to aim the baton at the base of the fire. Squeeze the trigger to spray the extinguisher. Continue spraying until the fire is completely out.

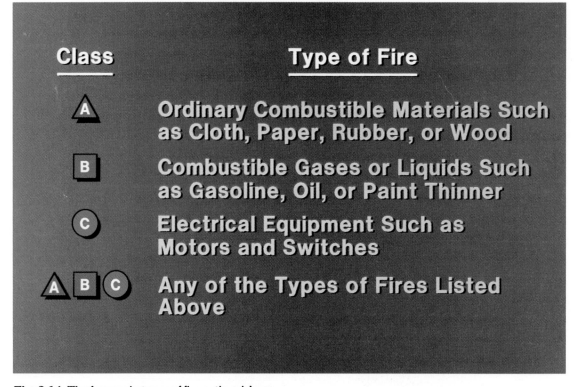

Class	Type of Fire
A	Ordinary Combustible Materials Such as Cloth, Paper, Rubber, or Wood
B	Combustible Gases or Liquids Such as Gasoline, Oil, or Paint Thinner
C	Electrical Equipment Such as Motors and Switches
A B C	Any of the Types of Fires Listed Above

Fig. 3-14. The four main types of fire extinguishers. All extinguishers are labeled to show the types of fires they can be used to put out.

Did You Know?

You may have heard of spontaneous combustion. This sometimes happens when oily rags are improperly stored. Basically, the combustible material in the rags joins with oxygen in the air. When this happens, small amounts of heat are created. As additional air joins with the combustible material, the temperature of the material rises. Finally, the rags burst into flame. Spontaneous combustion can be prevented by the exclusion of all air or by good ventilation.

First Aid

First aid is the immediate care given to a person who has been injured. The purpose of first aid is to temporarily relieve the pain caused by the injury or to protect the wound until further medical attention can be provided.

Being well trained in first aid procedures and having an adequately supplied first aid kit are essential ingredients in a safety program. Fig. 3-15. Everyone involved in construction should have a general knowledge of how to administer first aid. In addition, there should be at least one person present at all times who is well-trained in first aid.

A first aid kit that is fully equipped to handle emergency first aid needs should be located in a convenient place. Table 3-A provides a first aid table that explains the correct procedures to follow for several common industrial injuries.

Fig. 3-15. A well-stocked first aid cabinet is essential in any safety program.

Injury	Actions To Be Taken
Bleeding	Minor cuts: Wash cut with soap under warm running water. To stop bleeding, press hard with sterile compress directly over wound, then bandage. Severe cuts: Try to stop bleeding by pressing hard with a sterile compress directly over the cut. If bleeding continues or cut is deep, take patient to a medical facility.
Burns	Mild: Hold 2-3 minutes under cold running water or in ice water. If pain persists, apply petroleum jelly or mild burn ointment and bandage. Severe: Apply dry protective bandage. Do not try to clean burn or break blisters. Keep patient calm, and transport to a medical facility as soon as possible.
Electric Shock	Turn off electrical power if possible. If patient is still touching source of shock, do not approach until power has been turned off. If necessary, pull patient away from source of shock using rope, wooden pole, or loop of dry cloth. If breathing stops, start rescue breathing. Take patient to a medical facility.
Eye Irritation (chemical)	Pour water into corner of eye, letting it run to other side, until chemical is thoroughly flushed out. Cover with bandage and take patient to a medical facility.
Eye Irritation (foreign object)	Follow same procedure as above to flush particle out of eye. If object cannot be flushed and is visible in eye, touch lightly with corner of moist handkerchief. If object cannot be removed or seen, take patient to a medical facility. Do not allow patient to rub affected eye.

Table 3-A. This table shows basic first aid procedures. You should be familiar with these procedures in case an accident occurs in your laboratory.

For Discussion

1. Discuss the characteristics a person must have if he or she is to develop a good safety attitude.
2. Teamwork is an important element in safety. To ensure good teamwork, what characteristics must the people on a team have?

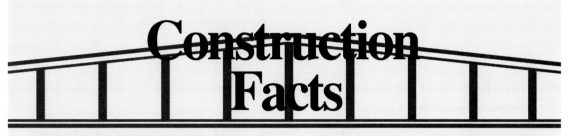

Construction Facts

WHEN THE PRICE OF A JOB WAS YOUR LIFE

In New York City in the spring of 1911, a fire broke out one Saturday in the Triangle Shirtwaist Company. The fire swept through the eighth floor of the manufacturing plant. One thousand people were working there at the time. When they tried to escape, they discovered the doors were locked. The fire-fighting equipment inside the building was totally inadequate. When help finally arrived, 146 workers were dead. Hundreds more were injured.

The Triangle fire called attention to the unsafe working conditions in our nation's industries. Work was often brutal, many times done by children. The laming of one little girl by a treadle machine nearly caused a strike. Her employer refused to find other work for her. Men working in mines and for the railroads usually had the most dangerous positions. A brakeman, for instance, had to stand between two railroad cars coming together. Then he dropped the pin that joined them. A brakeman with all ten fingers was usually recognized as new on the job. As one person put it, companies valued mules more than human beings. Finally, public sympathy was aroused.

Real change did not happen, however, until companies began to realize that accidents hurt them, too. Accidents halted production, raised costs, and created conditions that made more accidents likely. During the late 19th and early 20th centuries, a safety movement supported

by striking workers called for laws governing the workplace. Workmen's compensation laws stated that employers had to accept more responsibility for work injuries. Insurance and other health care programs came in later years.

Safety efforts paid off. Between 1926 and 1961, the rate of disabling injuries dropped from 32 percent to 6 percent. This drop in the accident rate was the result of reducing hazards and teaching workers accident prevention. Thanks to these and other safety measures, the industries of today are worlds apart from those of 100 years ago.

CHAPTER **3**

R E V I E W

Chapter Summary

Safety is a major concern in the construction industry. The primary concern of safety programs is to prevent injury to people and damage to property. To promote safety, the construction industry is careful to build structures that are safe to inhabit. They also promote worker safety programs. The United States government, insurance companies, and labor unions aid construction companies in promoting worker safety. A company safety program offers the best plan for accident prevention. In any safety program, good safety attitudes and general safety rules are important, as is fire prevention. Being well trained in first aid procedures and having an adequately supplied first aid kit is essential.

Test Your Knowledge

1. What five safety features can help protect the public near a construction site?
2. What are the two major safety concerns in the construction industry?
3. What distinguishes a worker with a good safety attitude?
4. Why are insurance companies concerned about safety on construction sites?
5. What should you do if you do not know how to operate a tool or machine?
6. Explain how heavy or bulky objects should be lifted and moved.
7. Why should you not carry tools in your pockets?
8. Why should you check to make sure three-prong electrical plugs are intact?
9. What type of fire extinguisher should be used to put out an electrical fire?
10. What is the purpose of first aid?

REVIEW

Activities

1. Design a safety poster that illustrates the proper use of a particular tool or machine. On the poster make a list of precautions that should be followed to use the tool or machine.

2. Make a list of accidents that could possibly occur in your laboratory. State why the accidents might happen and what could be done to prevent such accidents. Share your list with the class.

3. Write a report about an accident that happened in the construction industry. In the report, explain how the accident occurred and what happened to the worker(s). You may want to visit a local construction company, talk to a construction worker, or read a newspaper or industrial safety magazine to gather the information for this report.

Activity 1: Building a Studded Wall Model

Objective

After completing this activity, you should be able to demonstrate construction of a typical wall used in residential construction. This construction model will be to the scale of $1'' = 1'$.

Materials Needed

- $14 \frac{1}{8}'' \times \frac{1}{4}'' \times 8''$ strips of balsa wood
- 2 dozen $\frac{3}{4}''$ wire brads
- 1 small container of glue

Steps of Procedure

1. Lay two $\frac{1}{8}'' \times \frac{1}{4}'' \times 8''$ strips of balsa wood face to face. Be sure the ends are even. Refer to Fig. A.
2. Starting from one end, draw a line on the edges of each piece $1 \frac{5}{16}''$ apart. Including the line on each end, there should be seven of these lines. These lines will be the center marks for the wall studs. The piece of wood will be the top and bottom plates.
3. Nail and glue full studs ($\frac{1}{8}'' \times \frac{1}{4}'' \times 8''$) on all but the center marking. Fig. A.
4. Cut two strips of balsa wood to $6''$ lengths. Glue these to the inside of studs three and five. These pieces are called the trimmers. Fig. B.
5. Cut two $2 \frac{1}{2}''$ pieces of balsa wood. Glue them face to face. This will be the window header. Fig. B.
6. Glue and nail the header to the top of the trimmer and between studs three and five.
7. Cut three $2''$ and three $1 \frac{3}{4}''$ cripple studs from the $8''$ balsa wood. Fig. B.
8. Nail and glue the $1 \frac{3}{4}''$ cripple studs to the top plate and header at the center stud mark location, and to the inside of studs three and five. Fig. B.
9. Nail and glue the $2''$ cripple studs to the bottom plate at the center stud marking location, and to the inside of studs three and five. Fig. B.
10. Cut a window sill $2 \frac{1}{4}''$ in length.
11. Glue and nail the window sill to the tops of the cripple studs. Fig. A. You have now completed construction of the model wall.

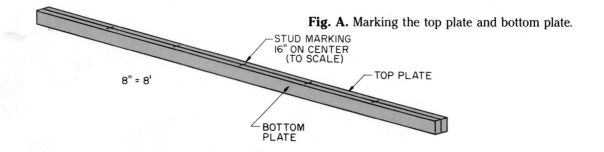

Fig. A. Marking the top plate and bottom plate.

STUD MARKING
16" ON CENTER
(TO SCALE)

TOP PLATE

8" = 8'

BOTTOM
PLATE

STUD WALL

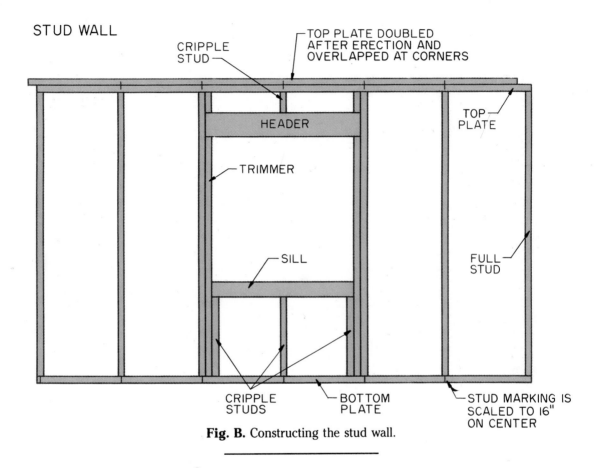

Fig. B. Constructing the stud wall.

Activity 2: Learning about Paint Color and Heat Absorption

Objective

After completing this activity, you will be familiar with the basic techniques of the finishing process of painting. You will have developed basic painting skills in either brushing or spraying. You also will be familiar with the principles of passive solar construction. Passive solar construction is the use of construction techniques to reduce energy losses. While developing the skill to paint, you also will be able to identify those paint colors that have greater heat absorption characteristics.

ACTIVITIES

Materials Needed
- spray or brush-on paint (black, white, green, red, yellow, blue)
- 2 dozen 1″-wide paintbrushes
- 1 shoebox per student
- 1 6″ × 16″ piece of cardboard
- 1 roll clear-plastic food wrap
- hot glue gun and glue
- 12 thermometers
- scribe
- paint thinner
- 1 dozen scissors

Steps of Procedure
1. Each student should bring to class a shoe-box with a lid. (A 6″ × 16″ piece of cardboard can be substituted for the lid.)
2. Your teacher will pass out the following materials to each student.
   - paint (in four colors)
   - 1 paintbrush
   - 15″ of clear-plastic food wrap
3. Your teacher will make the following materials available for the use of all class members.
   - hot glue and glue gun

Fig. A. Dividing the box into evenly-sized compartments.

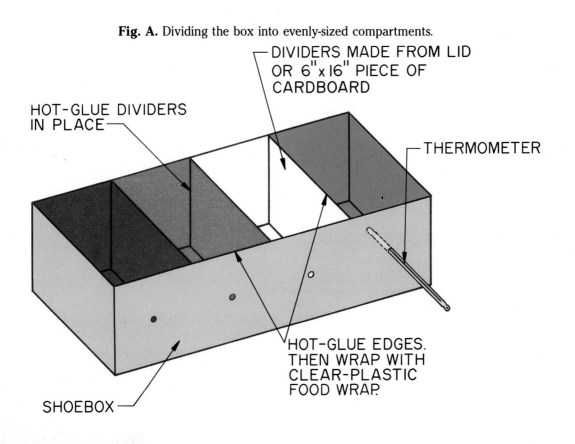

DIVIDERS MADE FROM LID OR 6"x16" PIECE OF CARDBOARD

HOT-GLUE DIVIDERS IN PLACE

THERMOMETER

HOT-GLUE EDGES. THEN WRAP WITH CLEAR-PLASTIC FOOD WRAP.

SHOEBOX

- thermometers
- scribes
- scissors
- paint thinner
- cleaning materials

4. Using the shoebox top or the 6″ × 16″ piece of cardboard, cut three pieces to match the height and width of the shoebox.

5. Hot-glue each of these pieces to the inside of the box to form four equal compartments. Fig. A.

6. Brush or spray each compartment a different color. Use all four colors.

7. After the paint has dried, turn the box on one side. Use a scribe to punch a hole in each compartment. Fig. A.

8. Coat the top edges of each compartment with hot glue. Coat the top of the divider and the side walls. Cover with plastic wrap. Press down the plastic wrap lightly so that it will be held by the glue.

9. Place each box right-side-up in sunlight or under a bright light. After 10 minutes, use a thermometer to check the temperature range in each compartment of the box. Insert the thermometer through the hole in each compartment.

10. What conclusions are you able to make from this experiment? Does paint color have an effect on the absorption of heat?

Activity 3: Describing Developing Technology

Objective

After completing this activity, you should be able to describe the basic development of a certain material, process, or tool in the area of construction technology.

Materials Needed
- plain paper (one sheet per student)
- 6 colored markers

Steps of Procedure

1. Divide a sheet of paper into four sections. Do this by first folding it in half from top to bottom. Then fold it in half from left to right. Unfold the paper. You have now divided the paper into four sections.

2. Brainstorm ideas for a construction-related topic. The topic may deal with any area of construction. For example, you might suggest ideas relating to the following topics:
   - materials
   - tools
   - safety
   - equipment
   - processes
   - structures

3. After choosing a particular topic, you should try to think of a "progression of development" that this item may have gone through. As an example of the general development of construction technology, you might refer to The Development of Construction Technol-

ACTIVITIES

72

Fig. A. The solution to one problem will sometimes create a new problem. Technology can be used to help solve problems.

DRAWING 1

People want to build a road between two villages. They find that the road will be blocked by a stream.

DRAWING 2

The people build a simple wooden bridge to span the stream.

DRAWING 3

The wooden bridge begins to crack beneath the weight of the wagon and its load.

DRAWING 4

To support the bridge, the people build a column of stones beneath the center of the bridge.

ogy, which begins on page 12. Each progression must show a problem that is being solved by using construction technology. Refer to Fig. A. Each progression should show:

- A problem (drawing 1).
- A solution through applied construction technology (drawing 2).

- A new problem related to and possibly caused by the solution (drawing 3).
- A solution for the new problem (drawing 4).

4. Sketch each development in one of the squares of your sheet of paper. You may use cartoon characters. You may even wish to include a little humor in your drawings. Write a caption for each drawing to explain what the drawing represents.

Activity 4: Making a Simple Electric Circuit

Objective

After completing this activity, you will have developed the basic skills needed to perform simple wiring activities. At the completion of this activity, you will be able to wire a plug, switch, light, and outlet into a circuit.

Materials Needed
- 1 wire cutter (side cutters)
- 1 wire stripper
- 1 screwdriver
- 6′ of #12 − 2 Romex wire with ground
- 3 wire nuts
- 1 three-prong plug
- 1 receptacle and receptacle box
- 1 single-pull, single-throw switch and switch box
- 1 junction box with light canopy and bulb
- 1 mounting board (a 2′ length of 2″ × 4″ will do)

- 6¼″ slotted screws
- 5½″ Romex box connectors

Steps of Procedure

There are three basic stages in this procedure: preparing the work station, running wires, and wiring the devices.

Preparing the Work Station

1. Cut the 6′ section of #12 − 2 Romex wire with ground into three equal sections. Each section should be 2′ in length.
2. Use the wire cutters to remove approximately 6″ of the wire coating. Be sure not to cut any of the internal wires or their coatings. Perform this step at both ends of each wire.
3. Use the wire strippers to remove approximately ¾″ of the wire insulation from the ends of each wire.

ACTIVITIES

4. Using two ¼ ″ screws, mount the receptacle box at one end of the 2 × 4. Now mount the switch box in the center. Mount the junction box at the remaining end of the 2 × 4.

Running Wires

1. Connect two Romex connectors to the receptacle box by placing one on each side. Tighten the lock-ring for each on the inside of the box.
2. Fasten a 2′ length of wire through each of the Romex connectors so that approximately 6″ of each wire remains inside the box.
3. Complete steps 1 and 2 once again for the switch box. Note that the feed line going into the switch box is the remaining end of the wire coming from the receptacle box.
4. Using the same method of fastening, connect the line coming out of the switch box to the junction box.

Wiring the Devices

1. Disassemble the plug. Insert the end of the wire through the plug cover. Tie an underwriter's knot 1 to 2 inches from the end of the wires. (Do not tie a knot in the ground wire.) This will keep the wire from being pulled from the end of the plug. See Fig. A.
2. Attach each wire to the appropriate terminal: black to brass, white to chrome, and ground to green.
3. Reassemble the plug. (Do not plug this in until all work has been completed and checked by your instructor.)
4. Next attach the receptacle. Fasten the black line coming into the receptacle box to a brass screw at the side of the receptacle. (Note

that for the receptacle and switching device there may be a strip gauge and hole to insert the wire.) Refer to Fig. B.

5. Now attach the remaining black line in the receptacle box to the remaining brass screw. This will allow a continuous flow of electricity to the switch box.
6. Attach the white wires to the chrome screws.
7. Attach the ground wire to the grounding terminal in the receptacle box and then to the green screw terminal of the receptacle. This will ground both the box and the receptacle.
8. To wire in the switch, fasten the black feed wire inside the switch box to one of the brass terminals at the side of the switch. Then fasten the remaining black wire to the remaining brass terminal on the switch.
9. Connect the white neutral wires by twisting them together with the square nose of the side cutters and capping with a wire nut. Be sure no portion of exposed wire is visible under the cap.
10. Fasten the grounds to the box and the switch.
11. Fasten a 6″ piece of black-and-white wire to your canopy light fixture. These are called wiring pigtails. Keep white to chrome and black to brass.
12. To complete the circuit, wire-nut the black wire from the junction box to the black pigtail from the light. Then connect the two remaining white wires and wire-nut together.
13. Ground the junction box.
14. Ask your instructor to check all work. Fig. B.
15. Fasten each device to each box.
16. Plug in the circuit. The outlet should be hot at all times. The switch will turn the light on and off.

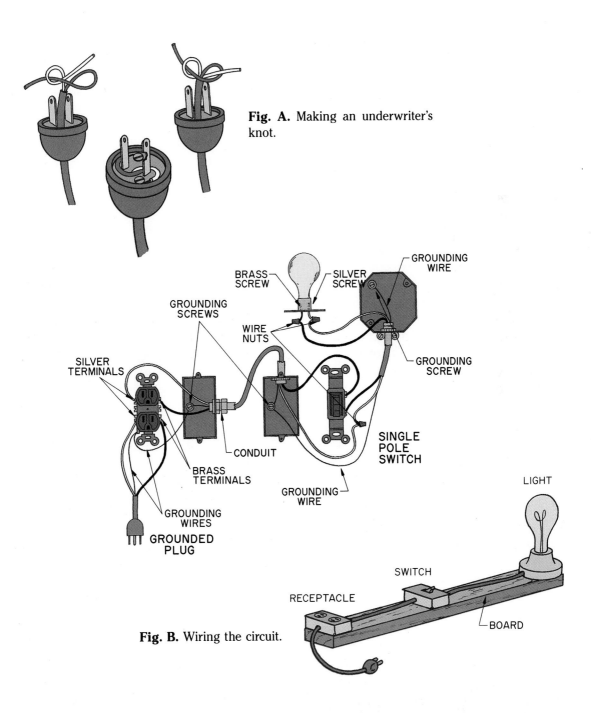

Fig. A. Making an underwriter's knot.

GROUNDING WIRE

BRASS SCREW

SILVER SCREW

GROUNDING SCREWS

WIRE NUTS

GROUNDING SCREW

SILVER TERMINALS

BRASS TERMINALS

CONDUIT

SINGLE POLE SWITCH

GROUNDING WIRES

GROUNDING WIRE

GROUNDED PLUG

LIGHT

SWITCH

RECEPTACLE

BOARD

Fig. B. Wiring the circuit.

76

SECTION

II MATERIALS, TOOLS, AND PROCESSES

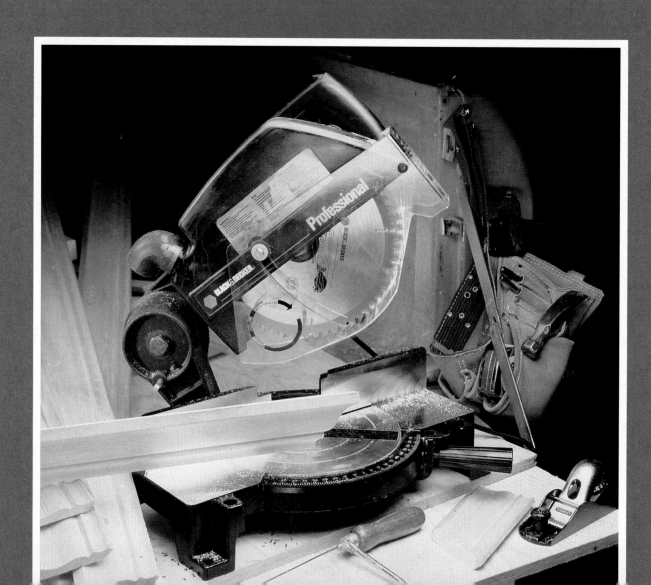

4 CONSTRUCTION MATERIALS

Terms to Know

adhesives	fiberboard	masonry cement	portland cement
admixtures	flooring	mesh	reinforcing bars
aggregate	hardboard	mortar	softwood
asphalt	hardwood	nominal size	standard stock
board lumber	insulation	nonferrous metals	structural steel
compressive strength	laminated beams	paneling	vapor barrier
concrete	laminated joists	particleboard	waferboard
contact cement	masonry	plywood	wood composites
dimension lumber			

Objectives

When you have completed reading this chapter, you should be able to do the following:

- List and identify the ingredients of concrete.
- Identify the types of wood and wood composites used in construction.
- Describe two kinds of masonry that are used in construction.
- Identify the uses of different kinds of metals in construction.
- Identify and describe materials used for insulation, interior surfaces, roofing, and flooring in construction.
- Identify several types of adhesives and mechanical fasteners and explain their uses.

The construction of a building requires hundreds of different materials. These materials must be chosen very carefully. The use of weak or inappropriate materials would result in an unsound and unsafe structure. It is therefore important that the people who work on a project know what materials are available. They must also know how to use each of the materials to the best advantage. For example, people who design and engineer a structure must make sure that the materials being used are appropriate for the project. The people who actually build the structure must use and install the materials correctly. This chapter will describe some common construction materials and will explain the uses of these materials. Fig. 4-1.

Fig. 4-1. This structure contains hundreds of different materials. The engineers and designers chose each type of material carefully to be sure it was the best choice for the project.

||||CONCRETE

Basically, **concrete** is a mixture of sand, rocks, and a binder. Concrete is one of the most common construction materials. Today nearly every structure contains some concrete. Concrete is used in building houses, skyscrapers, airports, and highways. Fig. 4-2. Even sewer piper can be made of concrete.

There are two types of concrete. *Asphaltic concrete* is a black concrete that is made with asphalt and rocks. This kind of concrete is used for paving roads and parking lots. Most often, however, the term *concrete* refers to *portland cement concrete*. The main ingredients in portland cement concrete are:
- Portland cement.
- Water.
- Aggregate.
- Admixtures.

Fig. 4-2. Here the basic materials in concrete are being blended.

Portland cement is the binder for concrete. Portland cement is a mixture of clay and limestone that has been roasted in a special oven called a *kiln*. The dried mixture is crushed and ground into a fine powder, as shown in Fig. 4-3. This powder hardens when it is mixed with water. As the portland cement hardens, it makes the whole mixture *adhere*, or stick together.

Water is mixed with the cement to cause the chemical reaction needed to harden the concrete. This water-cement mixture is called *cement paste*. The strength of the concrete depends on the amount of water that is used. Too much water dilutes the cement paste. This makes a weak concrete mixture. Too little water does not allow the proper chemical reaction to occur in the cement, which also causes the mixture to be weak.

Did You Know?

Portland cement was not named for Portland, Oregon. Portland cement was invented in 1824 in England. It was produced from a mixture of limestone and clay. The resulting product was called portland cement because some thought that it looked like portland stone. This was a limestone that was used for building in England. By 1850, the technique for producing portland cement had been improved. By 1875, portland cement was being manufactured in the United States.

Fig. 4-3. Portland cement is sold by the bag. Each bag weighs 94 pounds (42 kg).

Aggregate is the sand and rocks used in concrete. The main purpose of the aggregate is to take up space. About 75 percent of a batch of concrete is aggregate. Fig. 4-4. However, each piece of aggregate must be coated with cement paste. If the aggregate is not covered completely, the concrete will not be strong.

Different sizes of aggregate are needed to make a good batch of concrete. *Fine aggregate*, usually sand, is actually any aggregate that is smaller than ¼ inch in diameter. *Coarse aggregate* is any aggregate that is larger than ¼ inch in diameter. Aggregate can be gravel, crushed rocks, or slag. *Slag* is a by-product of steel manufacturing. At one time, slag was thrown away. Now it is commonly used for aggregate.

Admixtures are anything added to a batch of concrete other than cement, water, and aggregate. Admixtures are used to give concrete certain characteristics. For example, if brown concrete is wanted, an admixture consisting of brown coloring is added while the concrete is being mixed. A *chemical accelerator* is an admixture that makes the concrete set up fast. A *retarder* is an admixture that makes the concrete cure more slowly.

Concrete has a great amount of **compressive strength**, which means it can carry a lot of weight per square inch (psi). For example, a concrete driveway may be made from 3,000 psi concrete. When it has cured, this concrete will support 3,000 pounds per square inch.

Concrete does not cure at once. As it dries, or sets, it becomes hard. Fig. 4-5. However, it is not as strong as it will be when it has cured. Curing time is about one month. As it cures, the concrete becomes stronger. By the end of a month, the concrete has developed most of its strength.

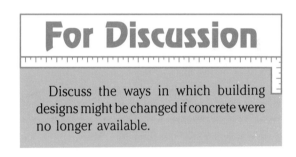

For Discussion

Discuss the ways in which building designs might be changed if concrete were no longer available.

Fig. 4-4. Aggregate mixed with portland cement.

Fig. 4-5. Concrete is poured into forms that hold it in place while it sets, or becomes hard. Forms give the concrete its final shape.

LUMBER AND WOOD COMPOSITES

Wood was one of the first materials to be used in construction. Although other materials are now used to meet many construction needs, wood remains one of the most popular and useful construction materials.

The wood used in construction is classified as either softwood or hardwood. Fig. 4-6. **Softwood** does not refer to wood that is soft; it refers to wood that comes from *coniferous* (evergreen) trees. Pine, fir, and spruce are some common softwoods that are used in construction. **Hardwood** is wood that comes from *deciduous* trees.

These trees shed their leaves each season. Oak, walnut, and maple are common types of hardwoods. Hardwoods are usually used for floors and for making cabinets, moldings, and other trim. Most of the lumber used in construction is softwood.

Lumber

Wood that is used in construction is called *lumber*. Two types of lumber that are commonly used in construction are dimension lumber and board lumber. **Dimension lumber** is lumber that measures between 2 and 5 inches thick. Dimension lumber is used to build framework for walls, floors, and roofs. Dimension lumber is classi-

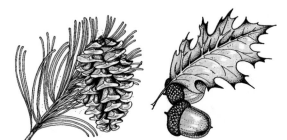

Fig. 4-6. Wood for construction comes from two types of trees. Softwood comes from coniferous, or evergreen trees. Hardwood comes from deciduous trees. Here, an evergreen is shown on the left. A hardwood is shown on the right. The inset above shows the needles and pine cone of the evergreen and the leaves and acorn of the oak.

fied by its size, or dimensions. For example, a piece of lumber that measures 2 inches by 4 inches is called a 2 × 4. Dimension lumber is generally available in even-numbered lengths, such as 8, 10, or 12 feet. Fig. 4-7.

Board lumber is lumber that measures less than 1½ inches thick and 4 or more inches wide. Boards are available in even-numbered widths, such as 2, 4, 6, 8, 10, and 12 inches. Like dimension lumber, all boards are generally available in even-numbered lengths.

It is important to realize that the stated size of lumber is not its actual finished size. For example, a 2 × 4 does not actually measure 2 inches by 4 inches. **Nominal size** is the size of the lumber when it is cut from the log. After it has been cut, the lumber is dried and then planed on all four sides to achieve smoothness. The finished size is therefore smaller than 2 inches by 4 inches. Table 4-A shows the nominal and actual sizes of softwood lumber.

Lumber is graded as it is cut from the log. The grade indicates the strength, usability, and appearance of the lumber. Table 4-B lists various grades of softwood lumber.

NOMINAL SIZE	ACTUAL SIZE
For Dimension Lumber:	
2 × 4	1½ × 3½
2 × 6	1½ × 5½
2 × 8	1½ × 7¼
2 × 10	1½ × 9¼
2 × 12	1½ × 11¼
For Board Lumber:	
1 × 4	¾ × 3½
1 × 6	¾ × 5½
1 × 8	¾ × 7¼
1 × 10	¾ × 9¼
1 × 12	¾ × 11¼

Table 4-A. The nominal and actual sizes of softwood lumber.

Wood Composites

Wood composites are those products that are made from a mixture of wood and other materials. Most wood composites are produced in large sheets, usually 4 feet wide and 8 feet long. Others, such as laminated wood joists and beams, are made to specification.

Plywood

One of the most commonly used wood composites is **plywood**. Plywood gets its name from its construction. It is made of several thin plies, or veneers, of wood that have been glued together. Each veneer is glued so that its grain is at right angles to the grain of the previous veneer. Fig. 4-8. The cross-layered grains make plywood very stable and strong.

Plywood is classified as *exterior*, for use outside, and *interior*, for use inside. The classification depends on the type of adhesive that is used to glue the veneers together. A waterproof glue must be used on exterior plywood. The glue that is used to make interior plywood is not waterproof, although interior plywood can withstand occasional moisture.

Fig. 4-7. Dimension lumber is used to frame walls, roofs, and floors.

BOARDS

APPEARANCE GRADES	Selects	B & BETTER (IWP — SUPREME) C SELECT (IWP — CHOICE) D SELECT (IWP — QUALITY)
	Finish	SUPERIOR PRIME E
	Paneling	CLEAR (ANY SELECT OR FINISH GRADE) NO. 2 COMMON SELECTED FOR KNOTTY PANELING NO. 3 COMMON SELECTED FOR KNOTTY PANELING
	Siding (Bevel Bungalow)	SUPERIOR PRIME

Table 4-B. Some common grades of softwood lumber.

DIMENSION/ALL SPECIES

Light Framing	CONSTRUCTION STANDARD UTILITY ECONOMY
Light Framing	SELECT STRUCTURAL NO. 1 NO. 2 NO. 3 ECONOMY
Appearance Framing	APPEARANCE
Structural Joists & Planks	SELECT STRUCTURAL NO. 1 NO. 2 NO. 3 ECONOMY
Decking	SELECTED DECKING COMMERCIAL DECKING
Studs	STUD ECONOMY STUD

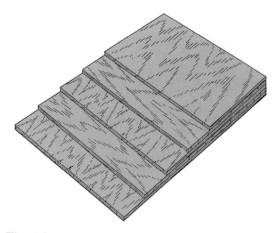

Fig. 4-8. Each ply of a piece of plywood is glued at right angles to the layers above and below it. This makes the plywood strong and resistant to warping.

Each side of a sheet of plywood is graded from *A* (highest grade) to *D*, according to the strength and appearance of its sides. Thus A-B plywood is plywood that has one excellent side and one side that has only a few minor flaws. Grades A and B are used when the plywood will be visible in the finished product. Grades C and D are used for general construction purposes. Table 4-C lists a few common grades and thicknesses of plywood.

Type	Description of Use	Grade of Plies		
		Face	Back	Inner Plies
Interior				
A–A INT	For use where both sides will be visible, such as for cabinets, furniture, and partitions	A	A	D
A–B INT	For use where two solid surfaces are needed, but the appearance of one of the sides does not have to be excellent	A	B	D
B–D INT	For use as backing and sides for built-in cabinets, where only one side must be sound	B	D	D
Exterior				
A–A EXT	For use on fences, signs, boats, and other items where both sides will be visible	A	A	C
B–C EXT	For use in buildings, truck linings, as base for exterior finishes for walls and roofs	B	C	C
MARINE EXT	For boat hulls and other items that will be exposed to water	A or B	A or B	B

Table 4-C. Some standard grades of plywood.

Other Wood Composites

Several other types of processed wood composites are also used in construction. Fig. 4-9. Each has special properties that make it useful in some aspect of construction. The following list describes some of the most commonly used wood composites. All of these products are sold in 4′ × 8′ sheets.

- **Particleboard** is made of small wood chips that have been pressed and glued together. Particleboard is commonly used in houses as an underlayment between the subfloor and the floor.
- **Waferboard** is made of large wood chips that are pressed and glued together and then cured with heat. Waferboard is not quite as strong as plywood, but it can be used instead of plywood in many applications.
- **Hardboard** is made up of very small, thread-like fibers of wood that are pressed together.

Because the fibers are so small, when they are pressed together they form a very smooth and hard material. The fibers are held together by *lignin*, a natural adhesive found in the fibers. Hardboard is sometimes called *Masonite®*, after the man who invented it.

- **Fiberboard** is made from vegetable fibers, such as corn or sugarcane stalks. It is not very strong, but it has good insulating properties and therefore is used as insulation sheathing beneath the exterior siding of buildings.
- **Paneling** is the term used to describe hardboard or plywood panels that have been prefinished. Paneling is used as a decorative finish on interior walls. It is available in a wide variety of colors, patterns, and wood grains.

Laminated Beams and Joists

Beams and *joists* are the horizontal framing, or supporting, members of a building. These

Fig. 4-9. Some common wood composites. Top to bottom: particleboard, waferboard, hardboard (Masonite®), fiberboard, and a prefinished panel.

parts must be strong, because they support the weight of other building materials. In many cases, dimension lumber can be used for beams and joists. However, sometimes a beam or joist must be larger than is available in dimension lumber. In other cases, a beam must be curved to fit a specific application. One solution to these problems is to use glue-laminated lumber, such as *laminated beams* and *laminated joists*.

Laminated beams are long, thin strips of wood that have been glued together. Fig. 4-10. If the beam is to be curved, the wood strips are bent around a form and clamped until the glue dries. This process is repeated until the desired shape and thickness are reached. The resulting beam is strong and durable.

Laminated joists are made of three parts: two *flanges* and a *web*. Fig. 4-11. The flanges are at right angles to the web, forming a cross-section that looks like the capital letter *I*. Each part is made of several plies of wood. A laminated joist is lighter than a dimension-lumber joist, but it is just as strong. Another advantage of a laminated joist is that it does not warp or shrink as easily as dimension lumber.

Fig. 4-10. The laminated beams support the roof in this building.

Fig. 4-11. This laminated floor joist is strong and lightweight.

Is wood still a common building material in your community? For what types of buildings is wood most commonly used? What do you suppose are the reasons for this?

MASONRY

Masonry is the process of using mortar to join bricks, blocks, or other units of construction. These materials are used mainly for exterior walls on houses and buildings. Masonry walls have several advantages over other types of walls. They are naturally fireproof and waterproof. They are also very durable and require little maintenance. Fig. 4-12. Bricks and concrete blocks are the most common types of masonry units, but there are several other types, such as stone and tile.

Bricks

Bricks are masonry units made from clay or shale. After the clay or shale has been dug from the ground, it is finely ground. Impurities are removed and the remaining material is dried. Then it is pressed into molds to form long, rectangular-shaped bars. When the bars are removed from the molds, they are cut into individual bricks. Then the bricks are fired, or baked, in a kiln. Firing causes a chemical change in the brick that makes it hard and strong.

Fig. 4-12. The brick exterior of this building requires very little maintenance.

Bricks are available in different sizes, shapes, textures, and colors. Fig. 4-13. Bricks may be solid or cored. *Cored* bricks are those that have holes. This reduces the weight of the bricks and provides an additional place to put mortar. Colors are determined by the type of clay or shale that is used and by the temperature of the firing process.

Did You Know?

In 1666, a great fire destroyed London. At that time, the city was built mostly of wood. As a consequence, the fire roared through the city, reducing more than four-fifths of London to ashes. After this fire, the city government drew up a plan for rebuilding the burned part of the city. To protect London against future destruction by fire, brick, rather than wood, was chosen as the main building construction material.

Concrete Blocks

Hollow concrete blocks are commonly used to construct walls. Concrete blocks have high compressive strength. This makes them especially good for use in walls that have to carry weight. Figure 4-14 shows typical shapes and sizes of concrete blocks.

Concrete blocks are made from a concrete mix that contains a small amount of water. The mix is placed into a block machine. In the block machine, the mix is squeezed and vibrated into molds to form the blocks. Most block machines can make about 15 blocks a minute. When the molded blocks emerge from the machine, they are placed in a kiln, where the concrete is cured. The temperature and humidity inside the kiln are carefully controlled, because the strength of the blocks depends on these factors.

Actual Dimensions			
	Width	**Height**	**Length**
Standard	3¾ "	2¼ "	8 "
Modular	3½ "	2⅙ "	7½ "
Roman	3½ "	1½ "	11½ "
Norman	3½ "	2⅙ "	11½ "
SCR	5½ "	2⅙ "	11½ "

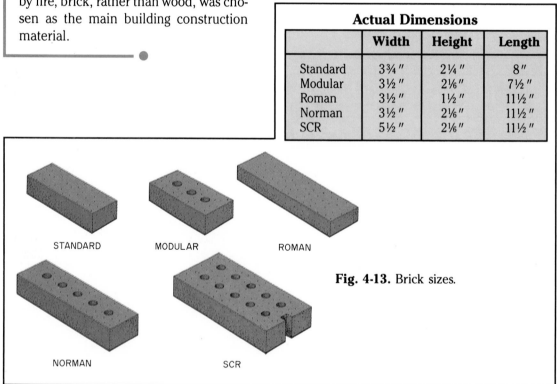

STANDARD   MODULAR   ROMAN

NORMAN   SCR

Fig. 4-13. Brick sizes.

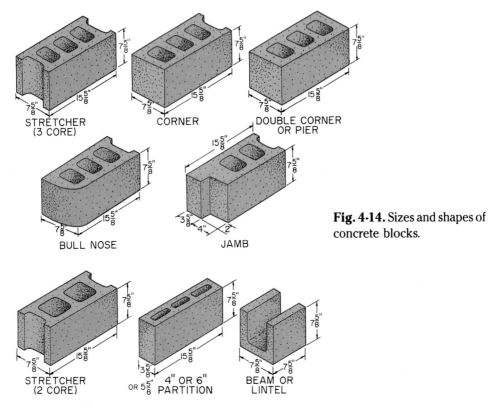

Fig. 4-14. Sizes and shapes of concrete blocks.

Mortar

All masonry units are held together with mortar. **Mortar** is a combination of masonry cement, sand, and water. **Masonry cement** is basically a commercially prepared mixture of portland cement and hydrated lime. The proportions of the ingredients in mortar vary according to its intended use. For example, the mortar used for an outside wall contains certain proportions of ingredients. Slightly different proportions of ingredients are used in the mortar for a wall that will not be exposed to water.

The mortar must be carefully mixed, because the strength of the finished product depends greatly on the strength of the mortar. If the wrong amounts of ingredients are used, or if the mixing time is too long or too short, the strength of the mortar is affected.

For Discussion

Have you ever seen stone walls in which the stones have been placed together without the use of mortar? How might the design of such a wall be different from the design of a wall in which mortar would be used? What difference would the use of mortar make in the way the stones could be placed?

▌▌▌METALS

Almost every structure in our society has some metal in its construction. Steel, an alloy of iron and other metals, is used primarily as structural reinforcement. Fig. 4-15. Other metals, such as aluminum and lead, are also used in construction. Each of these metals has special properties that are helpful in building houses, offices, and other structures.

Steel

In construction, steel is used primarily to provide support to structures. Some construction projects, such as radio, television, and electrical transmission towers, are made completely of steel. Steel is also commonly used in homes and other buildings.

Fig. 4-15. Steel beams and columns make up the framework of this building.

Structural Steel

Steel that is used to support any part of a structure is called **structural steel**. Horizontal steel *beams* and vertical steel *columns* can be fastened together to support walls, floors, and roofs. Structural steel can be processed into a number of different shapes and sizes. Fig. 4-16. These standard shapes and sizes are called **standard stock**.

Steel is classified according to the amount of carbon it contains. It becomes stronger and more brittle as its carbon content increases. Steel may contain from 0.15 to 0.65 percent carbon. Steel with a low percentage of carbon is used for buildings and bridges. High-carbon steel is used where the metal will be subject to extreme weather conditions or abrasion.

Special types of steel alloys have been developed to meet specific needs. For example, the outer layer of *weathering steel* is designed to rust. This layer of rust seals the steel from the atmosphere so that it cannot rust further. *Galvanized steel* is steel that has been coated with zinc to keep rusting to a minimum. *Stainless steel* has a chromium content high enough to allow chromium oxide to form on the outside of the steel. The chromium oxide protects the steel from rust.

Fig. 4-16. Common types of structural steel: I-beam, channel, angle, bars.

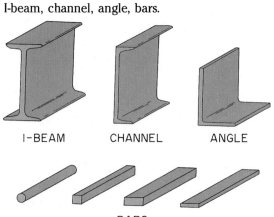

I-BEAM   CHANNEL   ANGLE

BARS

Fig. 4-17. Steel other than structural steel is used for many purposes in construction. These walkways are an example.

Bar Designation	Approximate Diameter (in inches)
#2	¼
#3	⅜
#4	½
#5	⅝
#6	¾
#7	⅞
#8	1
#9	1⅛
#10	1¼
#11	1⅜
#14	1¾
#18	2¼

Table 4-D. Sizes of reinforcing bars.

Did You Know?

The Eiffel Tower was constructed for the Paris Exposition of 1889. Nine hundred and eighty-four feet tall, the Eiffel Towel is nearly twice the height of the Great Pyramid. It was, however, built in just a few months. It was built at a low cost, using a small construction crew. The Eiffel Tower was an early and remarkable example of the uses of steel for building construction.

Miscellaneous Steel Parts

Besides structural steel parts, there are many other steel parts in a building. These miscellaneous steel parts include stairways, ladders, and open-grate flooring. Fig. 4-17. Low carbon content makes the steel workable so that it can be shaped more easily.

Steel-reinforced Concrete.

When concrete is used in a construction project, it is usually reinforced with steel. Steel makes the concrete less likely to break apart under stress. There are two methods of reinforcing with steel that are commonly used in construction. **Reinforcing bars**—or **re-bars**, for short—are steel bars that run through the inside of the concrete. Most reinforcing bars have ridges on them that help the concrete grip the bar. Table 4-D shows some sizes of the available reinforcing bars.

Another kind of reinforcing for concrete is called **mesh**. Made from steel wire, mesh looks like a wire fence. Fig. 4-18. The wire is welded together in two different directions to make it strong. Mesh is specified by the size of the openings and the size of the wire.

Nonferrous Metals

Metals that do not contain iron are called **nonferrous metals**. Because nonferrous metals do not rust, they have many applications in the construction of buildings. The nonferrous metals that are used most often are aluminum, lead, and copper.

Aluminum is a relatively lightweight metal that is very weather resistant. Some of its uses are for window frames, rain gutters, and siding on the outside of buildings. Some air-conditioning ducts also are made from aluminum sheet metal.

Lead is one of the heaviest and most dense metals. Because of its denseness, lead has the ability to deaden sound. Therefore, in some buildings sheets of lead are used for soundproofing. Lead is also used to cover the walls of X-ray rooms in hospitals and clinics. The lead shield keeps the harmful X-rays from escaping.

Copper is very malleable: it can be shaped easily. For this and other reasons, copper sheets are sometimes used on the roofs of buildings. Pipes for plumbing and electrical wire are also made from copper. Fig. 4-19.

For Discussion

1. Have you ever watched a building being constructed? How many of the steel items identified in this section have you seen used? Have you seen reinforcing bars (rebar) or mesh?
2. Are any buildings in your neighborhood made from weathering steel? What appearance do these buildings have?

Fig. 4-18. Wire mesh reinforcement is placed inside the form before concrete is placed.

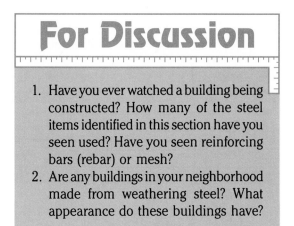

Fig. 4-19. Because copper can be shaped easily and is a good conductor, it is often used for electrical wire.

OTHER CONSTRUCTION MATERIALS

Insulation

Walls and ceilings are insulated to reduce the transfer of heat. **Insulation** helps keep heat from penetrating the building in summer and cold from penetrating in winter. Insulation is usually made from spun glass, foamed plastics, or certain vegetable and mineral fibers.

Insulation is available in many different forms. The form used depends largely on the shape and size of the space to be insulated. The four most commonly used forms of insulation are reflective, rigid, loose fill, and batt or blanket insulation.

HEALTH & SAFETY

Asbestos is a mineral fiber that occurs in nature. It is found in rock and is released by crushing the rock. For years, asbestos materials were used in building construction for a variety of purposes. Because asbestos is nonflammable, it was widely used as an insulating material. It has now been discovered that asbestos is a cancer-causing substance. Today, the use of asbestos in building construction is restricted.

Fig. 4-20. Sheets of insulating material are commonly used to sheath exterior walls.

Reflective Insulation

Reflective insulation is unique in that the material itself is not necessarily an insulator. Reflective insulation relies instead on a reflective foil surface. The foil reflects heat away from the structure being insulated. Reflective surfaces can also be placed on traditional forms of insulation to increase their insulating abilities.

Rigid Insulation

Natural fibers or plastic foam can be used to make rigid sheets of insulation. Rigid insulation is used to sheath, or cover, walls quickly and easily. Fig. 4-20. It can also be used to insulate many other flat surfaces. Rigid insulation is usually nailed or glued into place. Some rigid insulation has a reflective outer coating to increase its insulating properties.

Loose Fill Insulation

Loose fill can be made of fibrous or granular materials. Fibrous loose fill is made from fiberglass, wool, or vegetable fibers. Granular fill is made from granulated cork or from perlite or vermiculite. Perlite is a lightweight volcanic glass. Vermiculite is a lightweight material made from mica, which is a mineral.

Loose fill is usually blown into place through a special hose. It is used to insulate irregular surfaces that would otherwise be difficult to insulate. Fig. 4-21.

Batt or Blanket Insulation

The fibrous insulation materials contained in loose fill can also be formed into blankets and batts. *Blankets* are rolls of insulation designed to fit between framing members of a building. *Batts* are similar to blankets, except that they are not as long. Batts are made up to 8 feet (2.4 m) long, whereas blankets are available in rolls up to 24 feet (7.2 m) long. Batts and blankets are available in standard widths to fit between framing members set 16 or 24 (0.40 or 0.60 m) inches apart. Fig. 4-22. The insulation fits tightly so that pressure holds it in place. Sometimes the paper edge of the insulation is stapled to the wood to help hold it in place.

Fig. 4-22. Batts and blankets are designed to fit between framing members in wood-frame buildings.

Fig. 4-21. Loose fill insulation is used to insulate irregular surfaces.

Vapor Barrier

The types of insulation described in the previous section are designed to keep excessively cold or warm air out of a structure. These types of insulation are not effective against moisture, however. As the warm, moist air inside a building meets cold air from outside, the warm air cools. As it cools, its ability to hold moisture decreases. As a result, water condenses inside the walls of the building.

A special **vapor barrier** is needed to prevent the water from condensing. The vapor barrier is placed between the inside wall of the building and the insulation. This keeps the warm, moist air inside the house so that it does not meet the cold outside air. Fig. 4-23. The vapor barrier can be made of plastic, foil, or asphalt. It should always be placed on the interior side of the insulation.

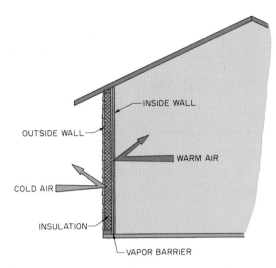
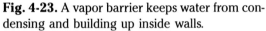

Fig. 4-23. A vapor barrier keeps water from condensing and building up inside walls.

Caulking

Although caulking is not in itself an insulating material, it does improve the performance of other insulating materials. Caulking is basically a gap-filling compound that is used to seal cracks and holes. Caulking helps to prevent unconditioned air from passing into or out of a home or building. A caulking gun is used to apply caulking around window trim, door trim, and other exterior and interior surfaces. Fig. 4-24.

Gypsum Wallboard

Gypsum wallboard is used to enclose interior walls and ceilings. This material is often called *sheet rock* or *drywall*. It is made of gypsum, a powdery mineral, sandwiched between sheets of special paper. The walls in most newer houses are covered with gypsum wallboard. Fig. 4-25.

Several types of gypsum wallboard are available. Regular wallboard has a smooth paper surface. After it is installed, it can be painted. Some wallboard is predecorated: a decorative finish is applied to it before it is installed. Fire-resistant wallboard is made with special additives that

Fig. 4-24. This person is using a caulking gun to apply caulking to cracks around window trim.

Fig. 4-25. Most interior walls are covered with gypsum wallboard.

make it able to withstand fire longer than regular wallboard. Moisture-resistant wallboard is used in bathrooms and other high-moisture areas.

Pipes and Wires

Most structures that are designed for human habitation have utility systems. Utility systems usually include plumbing, electrical, gas, and sewer systems. Pipes and wires are commonly used in utility systems to conduct fluids and electricity from one place to another. Without pipes and wires, electricity and indoor plumbing would be impossible.

Pipes

Water and other fluids are carried in pipes. Many different kinds of pipes have been developed to meet construction needs. Each kind has both advantages and disadvantages. Copper pipe, although expensive, is often used in plumbing because it does not rust or corrode and because it is flexible. Galvanized iron and cast iron are used because iron is very strong. However, iron is also very heavy and rigid. Clay pipe is used for outdoor sewer lines. There is even a special glass pipe that is used in some factories to carry acid.

Plastic pipes made of polyvinyl chloride (PVC) or polybutadene (PB) have the advantage of being lightweight, noncorrosive, and easy to work with. Fig. 4-26. However, plastic pipes have two major disadvantages: they cannot withstand high heat, and they cannot support as much weight as metal pipes.

Wires

Wires are used in the electrical system of a building. The most common type of electrical wire is called *nonmetallic sheathed cable*, or *romex*. It is usually made of copper and is sheathed with insulation. Wire sizes are shown in Fig. 4-27. Wires that may be exposed to mois-

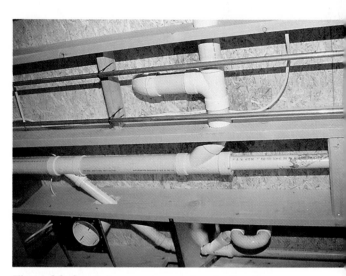

Fig. 4-26. Special connectors make plastic pipes easy to install.

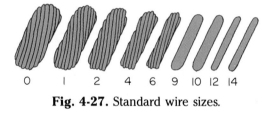

Fig. 4-27. Standard wire sizes.

ture or to a harsh environment are protected by conduits. A *conduit* is a metal or plastic tubing that encases wires to protect them.

Asphalt Roofing

Asphalt is a petroleum product made from crude oil. Four kinds of roofing materials are made from asphalt. *Coatings and cements* are liquids and pastes that are used to help seal the roof against moisture. A second kind of asphalt product is called *felt*. Felt is a clothlike material that is saturated with asphalt. It is commonly used as an underlayment on a roof to help provide protection from the weather. Other roof-

Fig. 4-28. Asphalt shingles are installed over felt underlayment.

ing materials are installed over the felt. *Roll roofing* is similar to felt, but roll roofing is thicker. It has a coating of fire-resistant mineral granules to provide protection from fire and to disperse the heat from the sun. The fourth asphalt product is *shingles*. Fig. 4-28. Shingles are made from the same material as roll roofing, but shingles are cut into short strips. For example, one common shingle size is 12 inches by 36 inches.

Did You Know?

The first sources of asphalt were natural sources. Asphalt occurs in certain lakes and springs. There, it is believed to be an early stage in the breakdown of small marine animals into petroleum. It is known that natural asphalt was used as early as 2000 B.C. It was used then to help make a water reservoir leakproof. Today, asphalt is manufactured from petroleum.

Flooring

Various kinds of **flooring**, or floor covering, are used in buildings. The floor covering that is used depends on the intended use of the area and the preference of the owner. Fig. 4-29.

Carpet is used in homes, stores, and offices to provide cushioning and an eye-pleasing effect. It is available in various styles and colors. The price range for carpet is wide enough to meet the needs of the majority of customers. Most carpet, however, shows wear quickly and must be replaced periodically.

Floor tile can be made of ceramic, asphalt, or vinyl. It is commonly used in areas where carpet is not practical. Floor tile is more easily cleaned and maintained than carpet and does not show wear as quickly. Some materials, such as vinyl, also come in 12-foot-wide sheets.

Wood floors are usually made of oak strips or squares and are very attractive. However, because oak flooring is expensive, it is not as commonly used as most other flooring materials. It is also relatively difficult to maintain.

Fig. 4-29. Different types of flooring. Clockwise from top left: carpet, floor tile, oak flooring, and terrazzo.

Terrazzo is a special kind of concrete floor that has marble chips embedded in it. After the concrete has cured, the surface is ground and polished. This decorative flooring is very durable. Liquid epoxy can be poured over the concrete floor to make it very tough.

Adhesives

Adhesives are materials that hold, or bond, other materials together. In order to do this, adhesives must have both adhesion and cohesion. *Adhesion* is the ability of one material to stick to another. *Cohesion* is the ability of the materials to stick together. Several types of adhesives are used in the construction industry. The most common are wood glue, contact cement, and mastic.

Wood Glue

For the most part, wood glue is used to glue wood materials together. There are many different types of wood glues that can be used in a wide variety of applications. Table 4-E lists the most common wood glues used in the construction industry.

Mastics

Mastics are thick, pastelike adhesives. They are often applied with a notched trowel or a caulking gun. Fig. 4-30. Mastics are very strong and often are used where it is difficult or inconvenient to use nails or other fasteners. For example, mastics are usually used to install wallboard, tile, and flooring materials over concrete or masonry surfaces.

Contact Cement

Contact cement is an adhesive that is applied to the surfaces of materials and then allowed to dry before the materials are combined. When the materials are finally combined, the contact cement instantly forms a permanent bond. No clamping is necessary. This makes

Table 4-E. Common wood glues used in construction.

Adhesive	Advantages	Disadvantages
Polyvinyl-Resin (White Glue)	Strong Inexpensive Fast drying	Low moisture and heat resistance
Aliphatic-Resin (Carpenter's Glue, Titebond)	Strong Heat resistant Slightly moisture resistant	Low moisture resistance
Urea-Resin	Moisture resistant	Not waterproof Must be mixed before using Low heat resistance Expensive
Recorcinol and Phenol Resourcinol-Resin	Waterproof Resists high temperatures	Must be mixed before using Leaves visible glue line
Epoxy-Resin	Water resistant Fills gaps	Must be mixed before using Expensive

Fig. 4-30. This construction worker is applying mastic in preparation for laying ceramic floor tile.

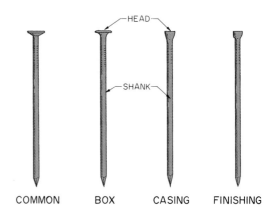

Fig. 4-31. Common, box, casing, and finishing nails are commonly used by construction workers. Notice the two main parts of the nail — the head and the shank.

contact cement a good choice for gluing large surfaces, where clamping would be difficult. Contact cement is used primarily for adhering plastic laminates to countertops and other surfaces.

Mechanical Fasteners

When you think of mechanical fasteners, you probably think of nails. However, there are also many other kinds of mechanical fasteners. Screws, nuts and bolts, and staples are a few of the most common. In this section you will learn more about the mechanical fasteners used in the construction industry.

Nails

Nails are probably the most commonly used mechanical fasteners in the construction industry. Many different types of nails are available. The most common are common nails, box nails, casing nails, and finishing nails. Fig. 4-31.

Nails vary both in length and in the diameter of their shanks. The word *penny* (abbreviated *d*) is used to describe the length of each type of nail. Because most nails are made from wire, the diameter of the shank is referred to by wire gage. Fig. 4-32.

Nails also vary in the materials from which they are made. Most nails are made of steel. However, because steel is susceptible to rusting, nails that will be exposed to moisture are *galvanized* (coated with zinc). Galvanized nails have much greater resistance to rusting than do non-galvanized, or *bright*, nails.

Common nails have large heads and diameters and are used mostly for rough framing work. *Box nails* are similar to common nails, except that box nails have slimmer shanks and thinner heads. Box nails are used for light framing work.

Casing nails and *finishing nails* are used mostly for trim work. They are less visible than common and box nails because of the shape of their heads. They are shaped so that the nail can be set beneath the surface of the wood with a nail set. A wood filler can then be applied over the nail to hide its location. Casing nails are larger than finishing nails and are used for heavier wood. Finishing nails are usually used for trim and molding on the interior of a building.

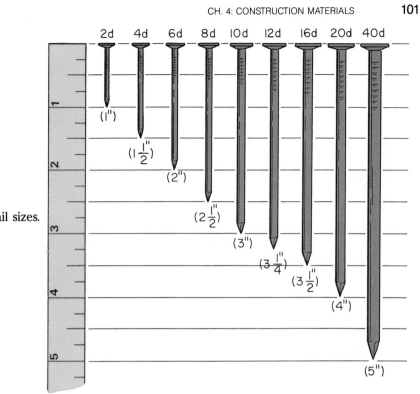

Fig. 4-32. Common nail sizes.

2d   4d   6d   8d   10d   12d   16d   20d   40d

(1") $(1\frac{1}{2}")$ (2") $(2\frac{1}{2}")$ (3") $(3\frac{1}{4}")$ $(3\frac{1}{2}")$ (4")   (5")

Many other special-purpose nails are also used in construction. For example, *scaffolding nails*, or *duplex-head nails*, are used for temporary structures, such as scaffolding and concrete forms. The double head on this kind of nail allows a worker to drive the nail into the wood to the first head. When the structure is disassembled, the nail can easily be pulled out using the second head. Special nails also have been developed for fastening drywall, flooring, rafters, shingles, and plywood. Fig. 4-33.

Screws

Screws are used to fasten wood, metal, and other materials together. *Wood screws* are designed to attach hardware or wood parts to wood. Although they are more difficult to use than nails, screws have more holding power.

They also allow the parts to be taken apart later without damaging the wood.

Drywall screws are used to attach wallboard or drywall to framing members. Drywall screws are installed with a special drill-like tool called a *power driver*. Fig. 4-34.

Like nails, screws are manufactured in many different lengths and diameters. The diameters are measured by gage numbers. The larger the gage number, the larger the diameter of the screw's shank.

Screws are also made with flat, round, or oval heads. Flat-head screws are designed to lay flush, or even, with the surface of the material. Round-head and oval-head screws are designed to extend above the surface of the material. Most screws are available with either a slotted head or a Phillips head. Fig. 4-35.

Scaffolding nail		Double head for easy removal from temporary wooden structures.
Plywood nail		Slim shank has rings to hold plywood better.
Scotch truss nail		Ribbed shank has extra holding power.
Roofing nail		Short, thick shank is designed to hold roofing material; large head helps seal the hole in the roofing material.
Specialty power nail		Shaped for use in power nailer.

Fig. 4-33. Specialty nails.

Fig. 4-34. A power driver and drywall screws.

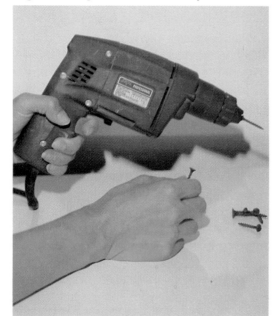

Fig. 4-35. Screws with flat, round, or oval heads are available with slotted or Phillips grips.

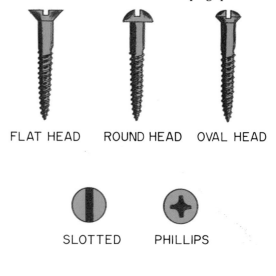

FLAT HEAD   ROUND HEAD   OVAL HEAD

SLOTTED   PHILLIPS

Other Fasteners

Many other types of mechanical fasteners are also used by construction workers. Three common types of bolts, for example, are shown in Fig. 4-36. *Carriage bolts* are used to fasten wood members together. *Machine bolts* are used to attach steel beams and other metal parts together. *Anchor bolts* are used to fasten wood plates to the top of concrete foundation walls. The bottom portion of the bolt is set in the concrete or in mortar in a concrete block. The top of the bolt is then fastened to the wood plate.

Framing anchors are used to strengthen framing joints. Fig. 4-37. A framing anchor consists of a piece of metal that has been bent to lay flat against both framing members. When a framing anchor has been nailed or bolted to both members, it helps support the joint.

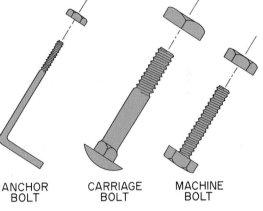

Fig. 4-36. Bolts commonly used in construction.

ANCHOR BOLT   CARRIAGE BOLT   MACHINE BOLT

Fig. 4-37. A framing anchor strengthens a joint between framing members.

Power fasteners are often used to reduce the time it takes to fasten framing and other construction materials. Fig. 4-38. *Power nailers* and *power staplers* use pneumatic force to insert nails and staples at the pull of a trigger. Most power nailers use either box nails or special nails designed especially for power fastening.

For Discussion

Which of the various types of insulation have you seen? What do you suppose has caused people to pay more attention to providing good insulation in their homes?

Fig. 4-38. Power fasteners allow workers to install mechanical fasteners quickly and easily.

Construction Facts

HOUSEWRAP

One of the biggest problems in making a house energy-efficient is controlling air infiltration. *Air infiltration* is the leakage of outside air into the air inside a house. Now a new material called *TYVEK® housewrap* can help control air infiltration.

TYVEK® housewrap is a flexible plastic sheet material. Made of very fine, high-density polyethylene fibers, it is very strong and will not rot. It is wrapped around the outside of the house just before the final siding or exterior covering is applied.

TYVEK® actually seals cracks and seams in the house. It blocks the air, yet allows the house to "breathe." In other words, moisture can escape, but there are no cracks for the wind to blow through. Reducing air infiltration can cut the cost of heating a home by approximately 25 percent. This can save homeowners a lot of money.

CHAPTER 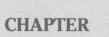 4

REVIEW

Chapter Summary

The construction of a building requires many different materials. They include concrete, which is a mixture of sand, rocks, and a binder. Wood is another material used in construction. For construction purposes, wood is classified as hardwood or softwood. Wood used in construction is called lumber. Wood composites are those products that are made from a mixture of wood and other materials. Wood composites include plywood, particleboard, waferboard, hardboard, fiberboard, and paneling. Masonry is the process of using mortar to join bricks, blocks, or other units of construction. Metals used in construction include steel, including structural steel and steel used to reinforce concrete. Metals that do not contain iron are called nonferrous metals. These include aluminum, lead, and copper. Insulation is used in a building to reduce the transfer of heat. Types of insulation include reflective insulation, rigid insulation, loose fill insulation, batt or blanket insulation, and vapor barrier. Gypsum wallboard, sometimes called sheet rock or drywall, is used to enclose interior walls and ceilings. Asphalt is a product commonly used in roofing products. There are various kinds of flooring, including carpet, floor tile, wood floors, and terrazzo. Adhesives are used to bond construction materials together. Mechanical fasteners include nails of various types and screws.

Test Your Knowledge

1. What are the main ingredients of portland cement concrete?
2. What is the main purpose of aggregate in concrete?
3. What is an admixture? Give an example.
4. What is the difference between softwood and hardwood?
5. Is most of the lumber used in construction made of softwood or hardwood?
6. Explain the difference between the nominal and actual sizes of a piece of lumber.
7. Name at least two types of wood composites.
8. Name two common types of masonry.
9. What is the purpose of mortar?
10. Name two types of steel reinforcers for concrete.
11. Name three common types of insulation material.
12. Name four kinds of flooring that are commonly used in buildings.

REVIEW

Activities

1. Take a close look at your house or apartment. Make a list of the materials mentioned in this chapter that you think were used to construct it. (Hints: Is your house made of wood, brick, or concrete block? How are the inside walls finished? Is there any paneling? What types of flooring were used?)

2. Visit a local do-it-yourself lumber supply store. With a ruler, measure the dimensions of 2 × 4s, 2 × 6s, 1 × 4s, 1 × 6s, and 1 × 10s. Compare your measurements with those listed in Fig. 5-8. What conclusion can you draw?

3. Collect wood shavings from a pencil sharpener. Use the shavings and wood glue to make a 2″ × 2″ sheet of particleboard. Be prepared to explain how to make the particleboard. (How did you mold it into a 2″ × 2″ sheet?)

4. Use thin balsa wood to create two sheets of plywood. Use wood glue to hold the plies together. In the first sheet, glue all of the plies with the grain in the same direction. In the second sheet, glue the plies at right angles, as described in this book. Test the strength of the two sheets. Be prepared to explain why plywood is constructed the way it is.

5 CONSTRUCTION TOOLS AND EQUIPMENT

Terms to Know

arc welding machines	crawler crane	nail set	scraper
backhoe	crosscut saw	pavers	sledgehammers
backsaw	digital rules	Phillips screwdriver	spiral ratchet screwdriver
blind riveter	equipment	pipe wrenches	standard screwdriver
brick trowel	excavator	pneumatic hammers	staplers
bulldozer	folding rules	portable circular saw	surveyor's level
bull float	framing square	powder-actuated stud driver	table saw
chalk line	front-end loaders	power drills	tape measures
claw hammer	grader	power miter saw	tower crane
cold chisel	hacksaw	power screwdriver	transit
compactor	hand tools	pry bars	trencher
concrete pump	heavy equipment	radial arm saw	truck crane
construction laser	laser-powered welder	ripsaw	twist drill bit
conveyor	level	rotary hammer	water pumps
cranes	miter gage	saber saw	wood chisels
	nailers		

Objectives

When you have completed reading this chapter, you should be able to do the following:

- List the four categories of construction tools and equipment, and give two examples of each.
- Recognize and know the uses of various hand tools and power tools.
- Recognize and know the uses of various kinds of construction equipment.
- Describe three pieces of heavy equipment.

Tools and equipment increase our ability to get a task done quickly and well. Tools and equipment are very important to the construction industry. Without them, construction projects would take a long time to complete, and some might never get done. Fig. 5-1.

You probably have heard the phrase "the right tool for the job." Each tool is designed to meet a certain need. For example, nails are driven by pounding them with a hammer. Screws are driven by twisting them with a screwdriver. You would cause damage and make little progress if you tried to pound a nail with a screwdriver or twist a screw with a hammer.

Sometimes a tool serves several purposes. For example, many hammers can be used both to drive nails and to pull nails. However, a tool should never be used to do a job for which it is not intended. Misusing a tool can result in damage to the tool, damage to the materials, and even injury to the person using the tool. For example, the delicate long-nosed pliers should not be used to apply force to a heavy object. Such abuse can easily bend the pliers' jaws out of line, making them useless for their intended purpose. Fig. 5-2.

The tools and equipment that are used in construction can be grouped into four categories:
- Hand tools
- Power tools
- Equipment
- Heavy equipment

Fig. 5-1. The interstate highway system is one project that might not have been attempted without today's advanced tools and equipment.

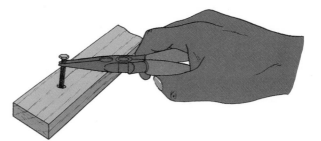

Fig. 5-2. Use long-nosed pliers and other tools only for their intended purposes. Misuse of tools can ruin them.

The construction industry requires many more tools than can possibly be described here. However, on the following pages you will find descriptions of some of the most common tools in each of the above categories.

HAND TOOLS

Hand tools are tools that use power supplied by a person. Most hand tools are relatively small and can be carried around by the worker. Many kinds of hand tools are used in construction. Some are very specialized and can be used only by trained workers. Others, such as hammers, saws, and screwdrivers, are very common. Carpenters, electricians, plumbers, and other workers use both common and specialized hand tools.

Measurement and Layout Tools

Accurate measurement and layout are among the most important tasks of construction because the rest of the construction process depends on them. No matter how well a structure is constructed, if it was laid out wrong, it will not meet the owner's requirements. Measurement and layout tools help surveyors lay out a site accurately. They also help carpenters and other workers build structures correctly according to the owner's specifications. Among the most common measurement and layout tools are those used for measuring distances, for squaring corners and framework, and for making straight horizontal and vertical lines.

Folding rules and **tape measures** are the most commonly used measurement tools. They are used to measure boards, pipe, wire, and many other construction materials. Tape measures are especially useful for measuring long or curved surfaces. **Digital rules** are used to measure relatively long distances, such as those in highway construction projects. Figs. 5-3, 5-4, and 5-5.

Another tool that is necessary to lay out a construction project is the **framing square**. Framing squares are used to measure 90-degree angles at the corners of framework and joints. They can also be used to measure cutting angles on dimension lumber. A framing square is made of a single piece of steel and marked with standard or metric units. Fig. 5-6.

A **level** is a long, straight tool that contains one or more vials of liquid. It is used to make

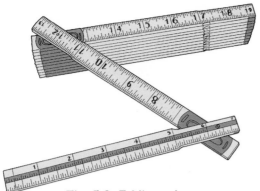

Fig. 5-3. Folding rule.

Fig. 5-6. Framing square.

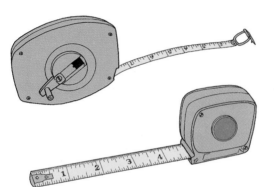

Fig. 5-4. Tape measures.

Fig. 5-5. Digital rule.

Fig. 5-7. Level.

sure that something is exactly horizontal (level), or vertical (plumb). When the level is exactly level or plumb, an air bubble in the liquid rests between two etched lines on each vial. Fig. 5-7.

Did You Know?

The bubble level used by carpenters today was introduced in 1661. At first, it was used on telescopes and on surveying instruments. It was not used by carpenters until the middle of the nineteenth century, when such levels began to be manufactured in quantity in factories.

A **chalk line**, or chalk box, is used to mark a straight line. A chalk line consists of a line (string) that is contained in a housing filled with chalk. The chalk coats the line as the line is drawn out of the housing. To use a chalk line, you fasten the hook at the end of the line to one end of the surface to be marked. Then you pull the line taut across the surface. You make the mark by lifting the line away from the surface and then allowing it to snap back. When the line hits the surface, it leaves a chalk tracing. Fig. 5-8.

Did You Know?

The chalk line has changed little in 5,000 years. It was used by the ancient Egyptians to mark a straight line between two points. The only real change has been in the color of the line itself. The ancient Egyptians colored the line with wet red or yellow ocher. Ocher is an iron ore that was used as a color pigment. The ancient Greeks however, colored the line with white and red chalk, in addition to wet ocher. Workers today follow the Greek practice.

Fig. 5-8. Chalk line.

Hammers

Hammers are tools that are used primarily for pounding. They come in many different shapes, sizes, and weights, according to their intended uses. One of the most common types of hammer is the **claw hammer**. The *face*, or pounding surface, of the claw hammer is used to drive nails. Opposite the face is a V-shaped notch called a *claw*. The claw is used to remove nails from boards. Most claw hammers have 16-ounce *heads*. Their handles are made of wood, steel, or fiberglass. The lower portion of some claw hammers is covered with a rubber grip to help keep the tool from slipping in the worker's hand. Fig. 5-9.

Sledgehammers are heavy hammers that are used to drive stakes into the ground and to break up concrete and stone. The head of a sledgehammer is made of steel and comes in standard weights of 5, 10, 15, or 20 pounds. The handle is usually made of sturdy hickory wood. Fig. 5-10.

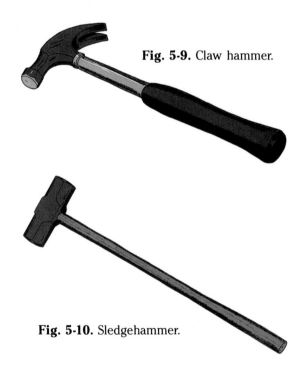

Fig. 5-9. Claw hammer.

Fig. 5-10. Sledgehammer.

Pry Bars

Carpenters use **pry bars** to pry the boards used to form concrete away from the concrete after it has set. Pry bars come in many different styles. One of the most common has a chisel at one end and a chisel with a claw for pulling nails at the other end. Fig. 5-11.

Screwdrivers

The two most common types of screwdrivers are standard and Phillips screwdrivers. The **standard screwdriver** has a flat tip and is designed to fit a standard slotted screw. The **Phillips screwdriver** has a tip shaped like an X. It is used to turn Phillips-head screws. The advantage of Phillips-head screws and screwdrivers is that the screwdriver grips the screw better. There is less chance of slipping. This reduces the chances of damaging the screwdriver, the screw, and the object being worked on. Figs. 5-12, 5-13.

A **spiral ratchet screwdriver** is one that relies on a pushing force rather than a twisting force. The tips for a spiral ratchet screwdriver are interchangeable, so it can be used for Phillips-head or standard slotted screws. The spiral ratchet screwdriver is faster to use than ordinary hand-twisted screwdrivers. It is therefore used when many screws must be driven in a short time. Fig. 5-14.

Saws

Various types of handsaws are used to cut lumber. Most handsaws look basically alike. Fig. 5-15. The main difference among them is the type of blade the saw has and the way it is used.

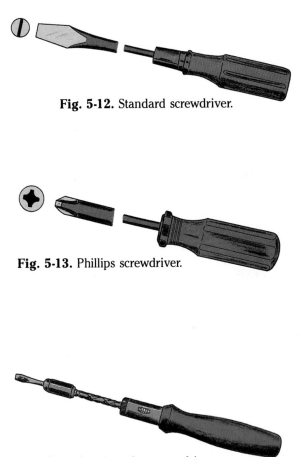

Fig. 5-12. Standard screwdriver.

Fig. 5-13. Phillips screwdriver.

Fig. 5-14. Spiral ratchet screwdriver.

Fig. 5-11. Pry bar.

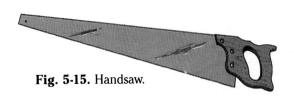

Fig. 5-15. Handsaw.

A **ripsaw** has chisel-like teeth designed for ripping, or cutting with the grain of the wood. Fig. 5-16.

A **crosscut saw** is used to cut across the grain of the wood. Its teeth are shaped and sharpened in such a way that it actually cuts two lines very close together and removes the sawdust between them. In this way it can achieve a smoother cut across the grain of the wood. Fig. 5-17.

A **backsaw** is a special kind of handsaw that has a very thin blade. The blade is reinforced with a heavy metal back called a *spine*. The spine keeps the thin saw blade from bending. As a result, the backsaw is used to make very straight cuts, such as those on trim and molding. Fig. 5-18.

A **hacksaw** is used to cut metal. Various types of hacksaw blades enable this saw to cut many different kinds of metal. Fine-tooth blades are used to saw thin sheet materials. Coarse-tooth blades are usually used for soft metals. Fig. 5-19.

Did You Know?

A saw is a blade with teeth. The saw developed over a period of thousands of years. There were a number of problems in obtaining a workable saw. The blade had to be thin, yet strong enough not to buckle. It was, perhaps, the Romans who devised the technique of alternating the teeth of a saw from one side to the other. This allows the sawdust to be pulled from the saw cut. The alternation of teeth also reduced the friction of the saw blade in the cut. This made sawing easier.

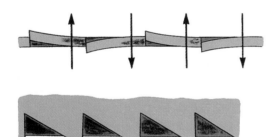

Fig. 5-16. Ripsaw blade. Top, top view. Bottom, side view.

Fig. 5-18. Backsaw.

Fig. 5-17. Crosscut-saw blade. Top, top view. Bottom, side view.

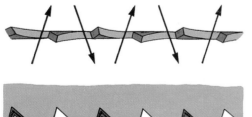

Fig. 5-19. Hacksaw.

Chisels

A chisel is a tool with a wedge-shaped blade that is used to separate materials. **Wood chisels** are used to trim wood. They are used to pare or clear away excess material from wood joints. They are also used to remove wood to make gains (recessed areas) for hinges. Wood chisels are usually powered only by human muscle, but a *mallet* can be used to provide more force if necessary. Fig. 5-20.

Metal objects can be cut with a **cold chisel**. Cold chisels are made of solid steel. They can be used to cut sheet metal, round objects such as chain links, bars, bolts, and various other types and shapes of metal. A cold chisel is driven by hammer blows to its flat end. Fig. 5-21.

Specialized Hand Tools

In addition to common hand tools that almost all workers use, each type of worker uses specialized tools. These tools are designed to help the worker do tasks that are peculiar to his or her career. For example, carpenters use a **nail set** to drive finishing nails below the surface of wooden trim and molding. They can then fill the resulting holes with a wood filler. When a finish is applied to the surface, the finish nails are not visible. Fig. 5-22.

Plumbers use **pipe wrenches** to turn objects that are round, such as pipes. The most commonly used pipe wrench is the *Stillson wrench*. A plumber can adjust a Stillson wrench to fit a wide variety of pipe sizes by turning an adjusting nut on the body of the wrench. Fig. 5-23.

One hand tool that is universally used by masons is the **brick trowel**. Masons use brick trowels to place and trim mortar between bricks or concrete blocks. Brick trowels are usually made of steel and have handles of wood or sturdy, high-impact plastic. Fig. 5-24.

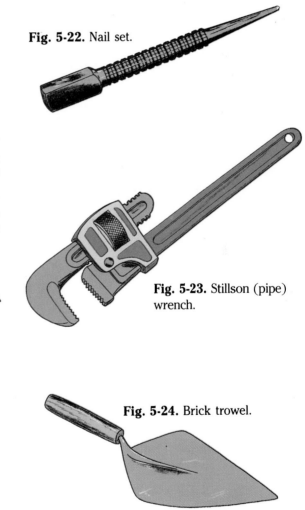

Fig. 5-22. Nail set.

Fig. 5-23. Stillson (pipe) wrench.

Fig. 5-20. Wood chisel.

Fig. 5-21. Cold chisel.

Fig. 5-24. Brick trowel.

Cement finishers use a **bull float** to smooth the surface of wet concrete. The face of the float, the part that touches the cement, is made of softwood. As the softwood is swept across the top of the wet cement, it leaves a smooth, slightly textured finish. Fig. 5-25.

Sheet metal workers use a **blind riveter** to fasten pieces of sheet metal together. The advantage of a blind riveter is that the worker can do the entire riveting operation from one side of the sheet metal. Fig. 5-26.

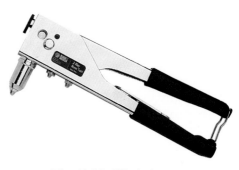

Fig. 5-26. Blind riveter.

For Discussion

Have you used any of the saws described in this section? Have you looked closely enough at the saws you use to see the way the teeth are arranged to cut effectively? Note that each tooth acts like a small chisel. How would this help in removing material?

Fig. 5-25. Bull float.

||| POWER TOOLS

Power tools are tools that are powered by forces other than human muscle. The power to operate most power tools is supplied by electric motors. A power tool usually provides more force than a person can. Therefore, with a power tool a worker can do a job faster and with less effort than by hand. Power tools often allow a worker to do a job more accurately, too.

The power that makes power tools so fast and efficient also makes them dangerous. Handle all power tools with respect. Be sure that you know how to use a tool *before* you use it. Follow all recommended safety precautions.

Power Saws

The many different cutting requirements for a construction project require several specialized types of saws. For example, a **radial arm saw** consists of a motor-driven saw blade that is hung on an arm over a table. This type of saw is used mostly for crosscutting 2×4's and other dimension lumber. It is also very good for cutting angles in rafters for roof frames. A radial arm saw is usually considered a *stationary* power tool. It is set up at one place on a construction site. The lumber is then brought there to be cut. Fig. 5-27.

Another stationary power saw is the table saw. A **table saw** consists of a blade mounted on an electric motor beneath a tablelike surface. Fig. 5-28. The blade sticks up through a slot in the table. The table saw is used for cutting large sheets of wood, plywood, and other wood products. It is also used for ripping lumber. The *fence*, or guide, is set at the desired distance from the blade. Once the fence has been locked in place, several pieces of lumber can be ripped to exactly the same width. Wood can be cut at an angle on a table saw by using an attachment called a **miter gage**. The miter gage can be adjusted to guide wood through the saw at an angle of up to 30 degrees. Portable miter saws also are available. Fig. 5-29.

Fig. 5-28. Table saw.

Fig. 5-27. Radial arm saw.

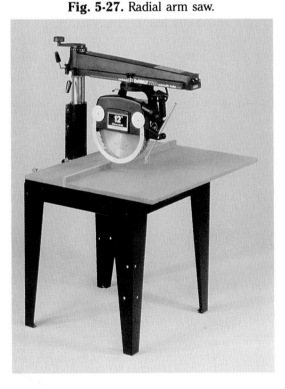

Fig. 5-29. Portable miter saw.

Construction workers also use several kinds of *portable* power saws. A **portable circular saw** is used to cut materials that are difficult to cut with stationary tools. Fig. 5-30. For example, a portable circular saw can be used to cut the ends off rafters after they have been assembled. A **power miter saw** is a circular saw mounted over a small table. Fig. 5-31. The saw pivots to enable the worker to cut various angles in wood.

A power miter saw is used to cut precise angles in wooden molding and trim. A **saber saw** has a small knife-shaped blade that reciprocates (moves up and down) to cut curves. Workers such as plumbers and carpenters use saber saws to cut holes in floors and roofs for pipes. Fig. 5-32.

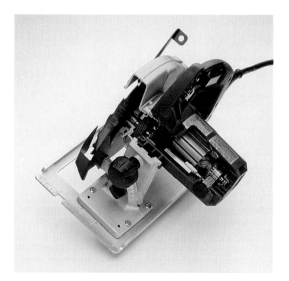

Fig. 5-30. Portable circular saw.

Fig. 5-31. Power miter saw.

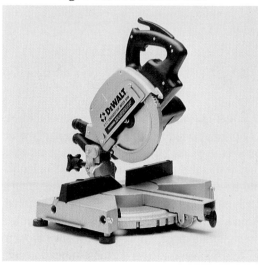

Power Drills and Screwdrivers

Power drills are used for drilling holes in wood, metal, and concrete. The size of a drill is determined by the chuck size and the power of the motor. The *chuck* is the part of a drill that holds the **twist drill bit**, or *bit*. A ½-inch drill will hold a bit of any size up to ½-inch in diameter. Some drills have reversible motors so that they can turn forward or backward. Some are even battery powered for use in places where there is no electrical service. Fig. 5-33.

A **power screwdriver**, or *screwgun*, is used to install and remove screws. It is similar to an electric drill. However, instead of a twist drill bit, a power screwdriver has a special screwdriver bit. The screwdriver bit fits into the head of the screw just like a screwdriver. Drywall installers use a power screwdriver to fasten wallboard to wall studs. Fig. 5-34.

Power Hammers

Power hammers are tools that strike with great force. **Pneumatic hammers**, or *jackhammers*, are used to break up concrete or asphalt paving. You have probably seen — and heard — one

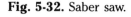

Fig. 5-32. Saber saw.

being used. An air compressor nearby supplies air at high pressure. A hose carries the air to the tool. The compressed air provides the power to move the bit, or chisel, up and down as it breaks into the paving. Fig. 5-35.

Another type of power hammer, the **rotary hammer**, looks like an electric drill. A rotary hammer operates with both rotating and reciprocating action. It is used to drill holes in concrete. Fig. 5-36.

Fig. 5-35. Pneumatic hammer.

Fig. 5-33. Power drill.

Fig. 5-34. Power screwdriver.

Fig. 5-36. Rotary hammer.

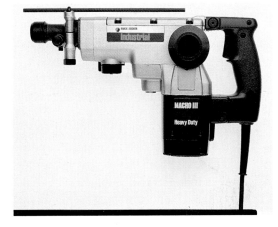

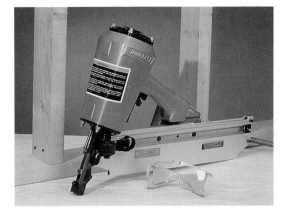

Fig. 5-37. Pneumatic nailer.

Power Nailers and Staplers

Nailers and staplers are power tools that fasten materials together. **Nailers** are sometimes called "nail guns" because they "shoot" nails. Most nailers are pneumatically powered. When the trigger of a nailer is pulled, air from an air compressor applies a strong force to the nail. Lumber can be nailed easily with a nailer just by aligning the boards and pulling the trigger. One pull on the trigger drives a nail all the way in. Fig. 5-37.

One specialized kind of nailer is a **powder-actuated stud driver**. It is powered by a 22-caliber cartridge that contains gunpowder. A special nail or fastener is put into the barrel, and the tool is loaded with a cartridge. The gun is pressed against the parts to be fastened, and the trigger is pulled. The powder-actuated stud driver can drive ½-inch- to 3-inch-long pins into wood, steel, or concrete. Carpenters use it to fasten 2 × 4's to concrete walls. It is also used to fasten metal door frames to concrete walls. Fig. 5-38.

Staplers work like nailers, but they are loaded with U-shaped staples instead of nails. Some are pneumatic, and others use electricity. Roofers use staplers to fasten roof shingles to decking. Carpenters use them to staple insulation into place. Fig. 5-39.

Fig. 5-38. Powder-actuated stud driver.

Fig. 5-39. Stapler.

For Discussion

The advantages of using portable power tools are obvious. Can you list a few ways in which the use of portable power tools has affected the construction industry.

||| EQUIPMENT

Equipment is a term that refers to large, complex tools and machines. Each type of equipment is designed to do a certain job. Most of the equipment that is used in construction falls into one of the following categories: surveying equipment, pumps, conveyors, and welding machines.

Surveying Equipment

Surveying equipment is used to measure land size and elevation. A survey is needed to begin almost any construction project. Surveyors mark off property boundaries and other important lines for construction workers. To do this, surveyors need accurate equipment.

A **transit** measures horizontal and vertical angles. Fig. 5-40. Surveyors use it to measure relative land elevation. They use a **surveyor's level** to find an unknown elevation from a known one. Fig. 5-41. A **construction laser** is a versatile instrument that can be used as a level or as an alignment tool. It flashes a narrow, accurate beam of light that workers can use as a baseline for additional measurements. Fig. 5-42.

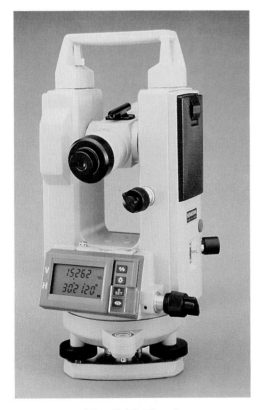

Fig. 5-40. Transit.

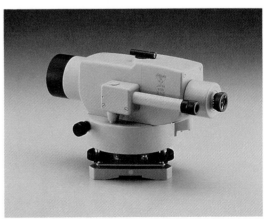

Fig. 5-41. Surveyor's level.

Fig. 5-42. Construction laser.

Pumps

The two kinds of pumps that are most commonly used in construction work are water pumps and concrete pumps. **Water pumps** are used to pump water out of holes in the ground so that work can be done. These pumps are usually powered by small gasoline engines. Because they do not require electricity, they are considered very portable. Fig. 5-43.

A **concrete pump** is used to move concrete from the concrete mixer to the concrete form efficiently. The pump is usually mounted on a truck. It has a *boom*, or long arm, that can be pointed in any direction. The boom holds and directs a hose through which the concrete is pumped. Fig. 5-44.

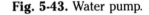

Fig. 5-43. Water pump.

Did You Know?

The Sumerians were an ancient Middle Eastern people. There is evidence that they practiced land surveying. Clay tablets from their culture record land measurements. Also, boundary stones used to mark land plots have been found. The ancient Egyptians also practiced land surveying. A drawing on the wall of a tomb at Thebes, in Egypt, shows two men measuring a field with what seems to be a rope marked at regular intervals. In fact, surveying may have begun in ancient Egypt. The Great Pyramid, which was built around 2700 BC, is accurately square. This indicates that the builders had a good knowledge of surveying.

Fig. 5-44. Concrete pump.

Conveyors

Materials other than fluids are often moved on a **conveyor**. You have probably seen conveyors in a supermarket or an airport. At the supermarket a conveyor automatically moves the groceries placed on it forward to the cashier. At the airport a conveyor moves luggage from ground level to a compartment in an airplane.

A similar type of conveyor is used in construction to speed the movement of materials. Roofers, for example, use conveyors to carry heavy bundles of shingles and other materials from the ground to the roof of a building. This allows the roofers to stay on the roof instead of spending time climbing up and down ladders. It also reduces the chances of workers being injured while climbing ladders. Fig. 5-45.

Fig. 5-45. The roof of this building covers over 3 acres (1.2 ha). In roofing such large buildings, contractors sometimes use conveyors to transport materials onto the roof.

Welding Machines

Ironworkers use **arc welding machines** to weld materials such as steel beams at construction sites. An arc welding machine, or *arc welder*, uses an electric arc to melt portions of the metal beams and thus the beams weld together. A gasoline engine powers the machine. The engine turns an electrical generator, which provides the electric arc. Fig. 5-46.

Another type of welding machine that is used in special situations is a **laser-powered welder**. A concentrated laser beam can heat metal to temperatures over 10,000 degrees Fahrenheit (5,540 °C). This makes the laser an ideal heat source for welding hard-to-melt metals, such as heat-resistant types of steel. Fig. 5-47.

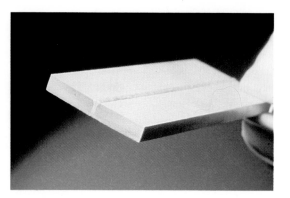

Fig. 5-47. The weld that joined two pieces of metal was made by a laser-powered welder.

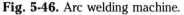

Fig. 5-46. Arc welding machine.

HEALTH & SAFETY

The use of any type of tool carries with it the risk of injury. In the United States, most states enacted some type of employer liability law between 1885 and 1910. Later, workmen's compensation laws were passed. These laws allow workers to recover damages when the disease or accident arose in the course of employment. Such laws now cover employment in most areas of business, not only in occupations commonly thought of as hazardous.

HEAVY EQUIPMENT

Some machines used in construction are very large and powerful. Such machines, called **heavy equipment**, are designed to do jobs that might be impossible to do by hand. Heavy equipment is expensive to buy and operate. It saves money in the long run, however, because it helps people do the work quickly and efficiently. Heavy equipment is used for jobs such as lifting and moving earth or other heavy materials. Can you imagine moving all the earth for an interstate highway using only shovels and wheelbarrows? That job would take a lifetime or more to complete without heavy equipment.

Heavy equipment includes machines such as cranes, excavators, bulldozers, and loaders, as well as equipment that is used in highway construction. There are several variations of each of these types of equipment. Each variation has been developed to meet specific needs on the construction site.

Cranes

Cranes are machines that lift large and heavy loads. They can also move loads horizontally by carrying them along a radius. Cranes are classified according to the weight they can lift safely. For example, a 50-ton crane can lift 50 tons or 100,000 pounds.

All cranes have similar basic parts. A *cable* is used to attach the object to be lifted to the hoist. The *hoist* is the mechanism that winds up the cable and does the lifting. The **boom** is the long arm of the crane that directs the cable. Several different kinds of tools can be put on the end of the cable to do specific jobs. The two most commonly used are the hook and the bucket. A *hook* is for general lifting. A *bucket* is used for digging and carrying. Fig. 5-48.

Most of the cranes in use today are *hydraulic*. These cranes have solid booms that telescope and extend hydraulically. *Mechanical* cranes are those with the familiar gridwork booms. Many different kinds of cranes have been developed for special purposes. Fig. 5-49. A **crawler crane** is mounted on metal treads so that it can move over rough terrain at a construction site. A **truck crane** is mounted on a truck frame so that it can be driven to the site.

A **tower crane**, or *climbing crane*, has a built-in jack that raises the crane from floor to floor as the building is constructed. A tower crane is used in the construction of tall buildings. It is usually positioned in the elevator shaft. When its job is done, another crane is used to remove the tower crane from the building.

Fig. 5-48. The main parts of cranes.

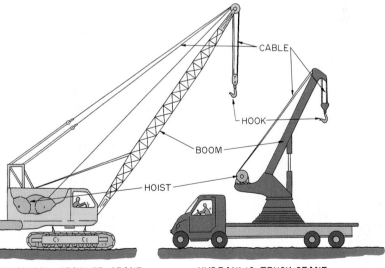

MECHANICAL CRAWLER CRANE   HYDRAULIC TRUCK CRANE

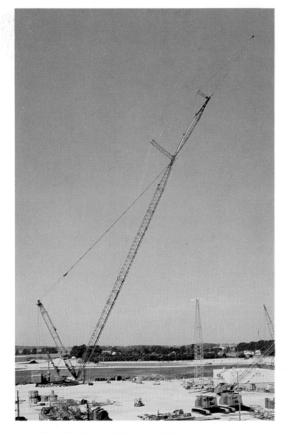

Fig. 5-49. A 640-foot (195-meter) crane.

Excavators

An **excavator** is a machine that is used for digging. It scoops up earth from one place and deposits it in another. Excavators are among the most common types of construction equipment because almost every construction job requires some excavation.

A **backhoe** is a type of excavator that is used for general digging. It is usually mounted on either a crawler or a truck frame. A dipper bucket is attached to a boom that is operated by hydraulic cylinders. The bucket is designed to dig toward the machine. The backhoe can dig below the surface on which it stands, so it is often used for digging holes. The size of the hole that is dug depends on the size of the bucket that is used. Fig. 5-50.

A **trencher** is a special kind of excavator that is used to dig *trenches*, or long, narrow ditches, for pipelines and cables. This machine is made of a series of small buckets attached to a wheel or chain. As the wheel or chain rotates, each bucket

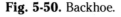

Fig. 5-50. Backhoe.

Fig. 5-51. Trencher-excavator.

digs a small amount of earth. The operator controls the depth and length of the trench. Trencher-excavators are also used for trenching. Fig. 5-51.

A *front-end loader* is sometimes used for excavation.

Front-end Loaders

Machines with large scoops used for shoveling are called **front-end loaders**. These machines are used to scoop up and deposit dirt or other materials. Loaders are often used to load trucks. Because a loader is mounted on a truck frame or a crawler, it can move small amounts of earth over short distances. Loaders have many other uses as well. For example, they are used to excavate basements, to backfill (fill in) ditches, and to lift and move construction materials. Front-end loaders are also called *scoop loaders* or *tractor shovels*. Fig. 5-52.

Fig. 5-52. Front-end loader.

Bulldozers

A tractor equipped with a front-mounted pushing blade is known as a **bulldozer**, or "dozer". The bulldozer is one of the most basic and versatile pieces of construction equipment. One of its primary parts is the blade, which is attached to the frame of the machine. Powerful hydraulic cylinders can move the blade up or down a short distance. Different blades are used for specific purposes. Many workers consider the straight blade to be the most useful. This blade is used to move earth over short distances. Another commonly used blade is slightly curved. It is designed for clearing land of bushes and trees. Fig. 5-53.

Most bulldozers are really crawler tractors. The steel tracks of the crawlers give excellent traction. This keeps the bulldozer from slipping when it pushes heavy loads. Other bulldozers have large rubber tires instead of crawlers. The rubber tires enable these machines to travel longer distances at higher speeds than crawler tractors.

Highway Construction Equipment

Highway construction requires a unique range of equipment, including some machines not used in any other form of construction. Most highway construction is done with four basic

Fig. 5-53. Bulldozer.

Fig. 5-54. Scraper.

kinds of machinery: scrapers, graders, compactors, and pavers. The many variations on these basic machines enable workers to adapt to local weather and soil conditions.

A **scraper** is a machine that is used for loading, hauling, and dumping soil over medium to long distances. Fig. 5-54. A **grader** is an earthworking machine that is used to grade, or level, the ground. It is used to prepare roadways and parking lots for paving. Fig. 5-55. A **compactor**, or *roller*, is used to compact the soil of a roadway just before the road is paved. Fig. 5-56.

Fig. 5-55. Grader.

Fig. 5-56. Compactor.

Types of compactors include steel drum rollers, tamping-foot rollers, grid or mesh rollers, and rubber-tired rollers. **Pavers** were developed to help in the construction of highways, parking lots, and airports. These machines place, spread, and finish concrete or asphalt paving material. Paving can be done very quickly using these machines. Fig. 5-57.

For Discussion

Are there any interstate highways near your city or town? What effect has the presence of such a highway had on building in your city or town, and especially near the interstate. Have you noticed that any particular type of building has been constructed?

Fig. 5-57. Paver.

JUMBO

The *jumbo* is a special piece of heavy equipment that was used in the construction of the Daniel Johnson Dam in Quebec, Canada. It is the world's largest hydraulic rock excavator, a monster machine weighing approximately 100 tons (91 metric tons). The jumbo has seven booms, each 12 feet (3.6 m) long. A hydraulic drill is located at the end of each boom. These drills were used to drill holes quickly in the hard Canadian granite. Dynamite was then placed in the holes to blast the rock apart. Because the jumbo could drill so many holes at the same time, the blasting work could be done much faster than with an ordinary excavator.

During the construction of the dam, extra care was taken to keep the jumbo in good working condition. Each day the machine was repaired and carefully maintained to prevent future problems. The repairs were made inside a specially built, heated garage for protection from the harsh Canadian winter with its −60 °F. (−51 °C.) temperatures. The preventive maintenance paid off. The machine did not lose a single day of work due to equipment breakdown—an impressive record for such a complex machine.

The Daniel Johnson Dam was completed faster and more accurately because of the jumbo. It was an expensive piece of heavy equipment, but the $2.5-million monster shortened a two-year project to just eight months. The jumbo was definitely the right tool for this job.

CHAPTER 5

REVIEW

Chapter Summary

Tools and equipment increase our ability to get a job done efficiently and quickly. Hand tools use power supplied by a person. These include measurement and layout tools, as well as hammers, pry bars, chisels, saws, and specialized hand tools. Power tools are tools powered by forces other than human muscle. Power tools include power saws, power drills and screwdrivers, power hammers, and power nailers and staplers. The term *equipment* refers to large complex tools and machines. These include surveying equipment, pumps, conveyors, and welding machines. Heavy equipment is designed for jobs that might be impossible to do by hand. Heavy equipment includes machines such as cranes, excavators, bulldozers, and loaders, as well as the equipment used in highway construction.

Test Your Knowledge

1. Name three problems that can occur when a tool is misused.
2. What are the four main categories of construction tools and equipment?
3. Name three common hand tools that are used by most workers on a construction site.
4. For what type of job can a backsaw be used?
5. What type of saw consists of a motor-driven saw blade that hangs on an arm over a table?
6. What two power tools mentioned in this chapter are considered stationary power tools?
7. Name two types of power tools that are used to fasten materials together.
8. What is meant by *equipment*?
9. Why are water pumps commonly needed on a construction site?
10. What type of welding machine can be used to weld heat-resistant metals? Why is this machine capable of welding such metals?
11. Name four types of heavy equipment that are used in construction.
12. Name four kinds of heavy equipment that are used specifically for road and parking lot construction.

REVIEW

Activities

1. Visit a nearby construction site. Observe the tools and equipment being used. Make a list of tools and equipment that you recognize from each category named in this chapter.

2. Look in a tool catalog to find five power tools. Then, for each power tool, find a hand tool that can do the same job. On a piece of paper, list each tool and its price. Then explain the advantages and disadvantages of using power tools and of using hand tools.

3. Go to a hardware store or lumberyard and identify five tools that are not described in this chapter. List and describe the uses of these tools on a piece of paper.

4. Select a career related to construction. Research the career and prepare a list of the tools and equipment that people in the career most commonly use. Research the career and prepare a brief report on how job skills and responsibilities may have changed over the last one hundred years.

5. Tools must be properly maintained and serviced. For example, a grinding wheel should be used to remove the mushroomed head of a cold chisel. Also, wood chisels should be kept sharp. Research the maintenance of small hand tools. Then prepare a short written report on common maintenance practices that will make hand tools more effective and safer to use.

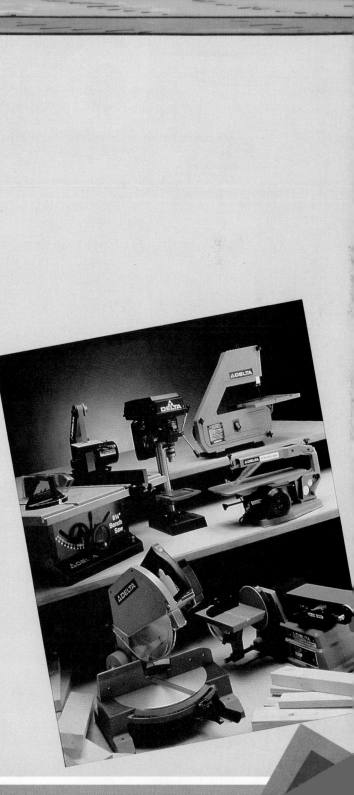

6

CONSTRUCTION FRONTIERS

Terms to Know

Com-ply®
computer-aided
   design (CAD)
geotextiles
materials
Micro-lam®

modular
   construction
module
oriented-strand
   board
prefabricated units

Objectives

When you have completed reading this chapter, you should be able to do the following:

- Describe several new construction materials.
- Describe several new construction methods.
- Describe how computers are being used in construction.

We live in a changing world. The goal of change is improvement—to be or do something better than before. Examples of change are all around us. Think of the changes in medical care and treatment in the last 50 years. Human heart transplants can now be done successfully. Even artificial hearts are in the realm of the possible. Think of the changes in transportation. Airplanes were invented less than 100 years ago, yet now we can fly almost anywhere in the world in less than one day.

The construction industry is changing, too. New materials and methods can help contractors do their work more quickly, accurately, and safely. Because of new materials and techniques, workers are becoming more productive.

Although no one can accurately predict the future, it is possible to make educated guesses about what could happen in the years ahead. By looking at recent developments, we can see the kinds of changes that might take place. This chapter will introduce you to some of the new materials and processes that are used in the construction industry today.

||||| NEW MATERIALS

Progress is made when people develop new, better materials to use and better ways in which to use them. **Materials** are the substances from which products are made. The materials used in construction must be durable, affordable, and easy to use. Some companies that manufacture building products hire researchers to search for and develop new materials. Their goal is to find better products and techniques that will make construction both safe and economical.

After an idea has been developed, the manufacturer must work to get it accepted by builders and by the general public. Several materials that were developed only recently are already being used in construction work. New types of adhesives, plastics, and wood composites are examples of such new materials.

Adhesives

An adhesive is any substance that can be used to *adhere*, or bond together, two objects. We think of adhesives in connection with paper, fabrics, glass, and, in some cases, wood. Now new, stronger adhesives are being used and perfected that can bond almost any material to almost any other material. Fig. 6-1. In construction, adhesives are being used more and more

Fig. 6-1. These concrete bridge segments are bonded together with an adhesive.

frequently instead of nails. Adhesives save time on a construction job. For example, drywall can be fastened to studs more quickly with adhesive than with nails.

Did You Know?

Adhesives have become important in construction. They are also important in joining pieces to make various objects of wood. Adhesives were used in ancient times. For example, some Egyptian carvings over 3,000 years old show a piece of veneer being glued to a wooden plank. In ancient times, glues were obtained from natural sources. The sticky resin obtained from some trees was used as a glue. Today, many glues are based on synthetic resins, which are products of the laboratory.

Plastics

Plastics have been used in many areas of construction for several years. New applications for plastics are being discovered almost daily. In fact, it is possible that plastics will someday replace most of the materials with which we now build. Some of the newer applications of plastics include products such as liquid storage tanks, roofing, and protective coatings. Waterproofing materials, engineering fabrics, and fasteners are also being made out of plastic.

Engineering fabrics are also called **geotextiles**. Geotextile material is like a large piece of plastic cloth. This fabric can be spread on the ground as an *underlayment*, or bottom layer, for roadbeds or slopes along a highway. Fig. 6-2. It is used to keep soil in place and to prevent erosion.

Structural shapes are another new application for plastics. Structural shapes made of fiberglass are now available for use where structural steel is not practical. For example, structural fiberglass can be used to build structures such as chemical plants, where steel would rust or corrode. Fig. 6-3.

Fig. 6-2. Geotextiles are used to stabilize soil. Here, geotextile is in place on a new road and paving is about to start.

Fig. 6-3. Fiberglass structural shapes can be used to build structures in hostile environments. Because of their resistance to rust and corrosion by salt water, fiberglass bolts are used on this tower.

Wood Materials

Even allowing for conservation and the replanting of trees, our timber supply is slowly being exhausted. Thus, it has become necessary to develop new ways of using our valuable wood resources. One of the first developments was plywood. Plywood has proved to be a popular and very useful building material. Other wood products are now being used in combination with wood and plywood. Some of the newest building materials made of wood are oriented-strand board, Com-ply®, and Micro-lam®.

Did You Know?

The Amazon river basin in Brazil contains the largest tropical rain forest in the world. Covering hundreds of thousands of square miles, it contains hundreds of varieties of trees. As an example, more than 100 species were counted in one-half a square mile. In recent years, vast areas of this forest have been cleared. Millions of trees have been destroyed and thousands of Indians displaced. The forest has been cleared to provide farmland and to harvest timber. However, as farmland, the land is poor. Very few trees are being planted to replace the timber that has been cut down. A rain forest, which takes thousands of years to develop, is being destroyed.

Oriented-strand Board

Oriented-strand board is made from small, crooked trees that otherwise would be unprofitable to harvest. About 60 percent of the product is made from pine. Hemlock and poplar make up the remaining 40 percent. All three woods are mixed together to make panels. Oriented-strand board can be used almost anywhere that plywood can be used.

Did You Know?

There are many types of plywood. Plywood is made from veneer. A veneer is a sheet of wood of even thickness. All plywood is made by gluing together a sheet of veneer to other sheets of veneer or to a solid wood core. It was the invention of modern wood glues that allowed the large-scale production of plywood.

Com-ply® and Micro-lam®

Com-ply®, or composit-ply, is made of several plies of veneer strips laminated to a core of particle board. **Micro-lam®** is made of pieces of veneer that have been laminated in a parallel direction. Fig. 6-4. Thin, dried veneer is coated with waterproof adhesive and bonded under heat and pressure. Micro-lam® is up to 30 percent stronger than comparable lumber. In addition, it uses 35 percent more of each tree than does regular lumber. The Micro-lam® process virtually eliminates warping, twisting, and shrinking.

New materials are not always accepted immediately. People tend to buy products they have used in the past—the ones they know will work. However, to meet the new and different construction needs of the future, builders, lawmakers, and consumers must sooner or later accept new ideas. To be sure that new materials are worthy of the people's trust, researchers test new materials extensively before putting them on the market. After a new product has been tested, building codes must be changed to include it so that contractors can legally use the material. This process takes time. However, new materials are usually approved and accepted by the people as the need for the new materials arises.

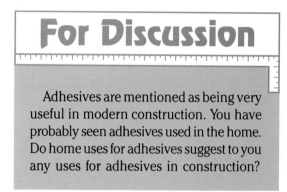

For Discussion

Adhesives are mentioned as being very useful in modern construction. You have probably seen adhesives used in the home. Do home uses for adhesives suggest to you any uses for adhesives in construction?

Fig. 6-4. This Micro-lam® beam was manufactured by gluing thin layers of wood together.

||| NEW
||| METHODS

New methods or processes of construction are developed mostly by construction engineers and by supply companies. Once developed, the new processes are tested thoroughly in laboratories before they are used in the construction industry. Many of these new methods have been approved and are now being used by more and more construction companies. Laser technology and prefabrication are two examples. Modular construction, another promising new construction technique, is still being developed in laboratory settings.

Laser Tools

Many uses have been found in the construction industry for lasers. The laser *emits*, or gives off, an absolutely straight beam of light. Its most obvious use is as an alignment tool: lasers are used to mark elevations in construction projects.

In addition to marking elevations, lasers can help construction crews maintain elevations. For example, a laser is mounted on a tripod at the construction site. An electronic receiver is then attached to the blade of a grader. The laser automatically controls the height of the grader blade. This enables the grader to achieve an exact elevation. Fig. 6-5. Lasers can guide paving machines and trenching machines in a similar way.

A concentrated laser beam can reach very high temperatures. Therefore, lasers can be used to weld heat-resistant metals. For example, some types of structural steel can benefit from laser welding techniques. On a similar principle, lasers can be used to drill holes in very hard surfaces.

HEALTH & SAFETY

Some lasers are extremely powerful. These lasers can produce serious injury. Most lasers, however, are low-power. Such lasers do not produce a light as intense as that produced by more powerful lasers. Even with these lasers, however, you should not look directly into the laser beam or into any reflection of the laser beam.

Fig. 6-5. The electronic receivers on this grader blade receive a signal from the laser. The laser controls the height of the blade, making the correct elevation easier to accomplish.

Prefabrication

Components of structures and even whole structures are now being built in factories and shipped to the building site. Fig. 6-6. Such buildings are called **prefabricated units**. The units are shipped complete. They include the trim, plumbing, insulation, doors, and even molded-plastic bathrooms. The units are assembled at the construction site. This method is a quick and efficient way to construct a building. Fig. 6-7.

Modular Construction

Another method of construction that falls somewhere between prefabrication and traditional construction is called **modular construction**. In this method, a building is designed to be constructed with modules. A **module** is simply a standard unit that has been chosen by a manufacturer, such as 4 inches. Modules allow the material supplier to stock pieces in standard sizes. When the building is to be con-

Fig. 6-6. This factory manufactures prefabricated housing units. These units are delivered to the site almost complete.

structed, the contractor orders all the parts, which have already been cut to size in the factory. The builder needs only to assemble the pieces. Little or no cutting is needed at the site.

Did You Know?

Large stone blocks were used to build the pyramids of ancient Egypt. Archeologists have found evidence that indicates that these large blocks were cut to roughly the needed shape at the quarry. They were then cut to exact size just before they were placed on the pyramid. Strictly speaking, this was not prefabrication. It was, however, precutting, which is a step in prefabrication.

Fig. 6-7. This motel was made from prefabricated units.

For Discussion

You may have seen prefabricated buildings being constructed in your city or town. What advantages do you think there are to constructing a prefabricated building? Can you think of any disadvantages?

‖‖‖ COMPUTERS

Computers are now found in most businesses, and construction is no exception. Computers are used for both designing and engineering. Sometimes computers are even used to help run construction companies.

Computer-aided Design

The use of computers for designing and engineering construction projects has become common. **Computer-aided design (CAD)** speeds up the designing and engineering work of a project. CAD also provides great accuracy in design calculations. Fig. 6-8.

The drawings and specifications for a project can be made easily and quickly using a computer. A CAD graphics program connected to a plotter can be programmed to draw a plan in just a few minutes. Specifications are prepared using a word processing program and a printer.

Computers and Construction Management

Computers are used not only for design and engineering, but also in the management of the construction firm. Computers are used for estimating the costs of a project. They are also used to schedule the many tasks involved in a project and to keep track of the schedule. Fig. 6-9. Computers are used to keep track of material and labor costs. Of course, computers are also used for the recordkeeping tasks that are part of any business. Such tasks include payroll, inventory, and billing.

By using computers, construction managers are able to receive detailed reports of every phase of a project in a short time. Project reports help managers find opportunities to improve productivity at the job site. Using a computer helps a contractor stay in control of costs and scheduling. In effect, the computer helps a construction company compete more effectively with other firms.

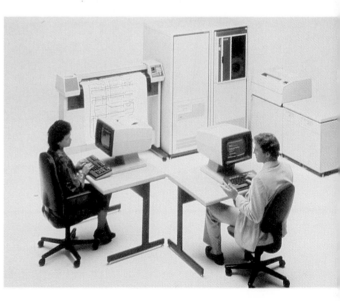

Fig. 6-8. A plotter connected to a computer can draw plans very quickly.

Fig. 6-9. A computer can help managers keep projects on schedule.

THE FUTURE

What will the future hold for construction? There will be changes in every area, from the appearance of structures to the skills of the people that build them. New equipment will be developed that can do jobs more quickly and efficiently. Equipment will be more powerful and efficient than that which is used today. More tools will be created to do specific jobs. The need for unskilled labor may decrease. Instead, there will be a need for people who know specific skills. New materials will create a need for people specially trained to handle them. Some workers will have to be retrained as new materials and processes are introduced.

Space Habitats

In the future, a construction crew may be called upon to build a space station several hundred miles into space. Fig. 6-10. Such a structure would have to be self-contained. It would have to have its own atmosphere, temperature control system, and, perhaps, a gravity system.

Such a space station would be known as a space *habitat*. As we continue to expand our horizons, we will have to create new designs and building techniques to meet special needs. One element, however, will remain the same: construction will continue to be the interesting challenging industry it is today.

For Discussion

The construction of a space station would face certain problems. Among the problems would be that of working in an environment without gravity. How would such an environment affect work practices? What unusual precautions would need to be taken?

Fig. 6-10. Someday structures may be built in space.

Construction Facts

INTELLIGENT BUILDINGS

New electronic technology has allowed buildings to become "intelligent." Building intelligence refers to the capabilities of electronic equipment and systems that are built into a building.

The basic idea of a smart building is that a network of sensors gathers data about the building environment. A computer uses the information from the sensors to adjust the building controls to meet changing conditions. Such a system can control the mechanical, fire and life safety, security, and energy management systems. It can also control elevators, on-site data communications, and telecommunications.

One example is a system that saves energy by sensing occupancy of rooms in the building. In this sytem, special sensors mounted in the ceiling are wired to the lights. When a sensor detects human body heat, it causes the lights in that room to turn on. The lights turn off 12 minutes after a room has been vacated. This system can reduce a monthly electric bill by about half.

Smart buildings are becoming more and more popular. Many existing buildings are being retrofitted for high-tech electronic systems that will make them smarter. Electronic systems are also easy to include in designs for new buildings. Research is now being done on other electronic systems that may make buildings even smarter in the future.

ACTIVITIES

Activity 1: Measuring an Incline Using an Inclinometer

Objective

After completing this activity, you will know how to construct and use a simple inclinometer. This can be used to measure the degree of slope for a driveway, patio, or other land forms.

Materials Needed
- 10 sheets of grid paper (10 squares per inch)
- 1 piece of ⅝″ × 24″ × 36″ plywood
- 1 piece of 22″ × 28″ white posterboard
- 1 piece of strong nylon fishing line 24″ long
- 1 lead fishing weight with attachment eye
- 1 small, clear protractor
- 1 wood screw, 1″ long

Your teacher will guide you through the construction and use of your inclinometer. He or she also will provide the following items:
- adhesive to attach posterboard to plywood
- overhead projector
- fine-point permanent marker
- 6′ tape measure
- 18″ metal straightedge
- screwdriver, level, scissors, and other necessary tools.

Steps of Procedure
1. Your first step is to construct a large cardboard protractor. This can be done by projecting a clear protractor onto a 22″ × 28″ piece of white posterboard with an overhead projector. The projected protractor should not be larger than 26″ wide and 13″ high.

2. Accurately trace the shape of the protractor and the individual degree markings with a pencil. Go over the markings with a fine-point permanent marker. Label the degree markings as shown in Fig. A. Next, trim the piece of posterboard so it will easily fit onto the 24″ × 36″ piece of plywood.

3. Now measure over 18″ from one end of the long side of the plywood board. Use a carpenter's square to mark a straight line down from the top, passing through the 18″ mark. This represents the middle of the board. Measure 2″ down from the top of the board. Place a mark on the line you have just drawn. This locates the position where you will place the 1″ wood screw. In placing the screw, make certain it is straight and not turned all the way through the plywood.

4. Wrap one end of the nylon fishing line around the projecting portion of the wood screw. Tie the line securely. Place the other end of the line through the eye of the fishing weight. Adjust the length of the line so the weight is at least 3″ above the bottom edge of the board. Tie the weight securely. Cut off the excess line.

5. Use a carpenter's level to find a flat surface that is exactly level. Now place the bottom edge of the plywood on this surface. Allow it to remain in this position until the fishing weight stops moving. Now mark the exact position of the line on the board just above the weight. Use a metal straightedge and pen-

Fig. A. Constructing and making the inclinometer.

cil to extend this mark up to the center point of the wood screw.

6. Next, place adhesive on the back of the protractor. Position the protractor on the board so the center mark along the straight portion of the protractor is against the midpoint of the wood screw. The 0° mark should be placed exactly on the line you have just marked near the bottom of the board.

7. You can also make your inclinometer into a measuring board by creating a simple ruler along the bottom edge. Refer to Fig. A to see how the board is marked with a fine-point permanent marker. This scale will be useful when you begin using your inclinometer.

8. You are now ready to measure the incline of one or more driveways in your neighborhood. Beginning at the street, with the scale facing you, place the left end of the board at the very end of the drive. The weight will swing to a position on the protractor. If the drive slopes uphill, the line will be to the

ACTIVITIES

left of the 0° mark. If the drive slopes downhill, the line will be to the right of the 0° mark.

9. When the line stops moving, read and record the degree of incline. This represents the degree of incline in the first three feet of the driveway. Move the left end of the board forward to the point at which the right end was located for the first measurement. Once again, read and record the degree of incline. Repeat this procedure for the entire length of the driveway. Fig. B.

10. The data you have collected can be graphically shown on a sheet of graph paper. Begin by drawing a straight line the length of your grid paper. Let four squares represent one foot. Place a mark every three feet along the line. Now place the protractor on the pencil line with the midpoint on the beginning of your measurement. Next, refer to the data to see what the first degree of incline was. In the example provided it was 12°. Place a small mark to indicate the 12° position.

11. Use a straightedge to project this back toward the beginning point and directly over the 3″ mark. Then move the protractor and locate its midpoint directly on the mark you have just made. If there was an incline in the first 3″, this means your protractor will be slightly above the straight line you began with. Repeat this process using all the data you have collected.

12. What would you conclude if you discovered the incline of the driveway sloped toward the garage door?

Other Applications

1. Patios should slope away from the house and not have small dips. This will avoid water damage to the house or standing pools of water. The inclinometer can be used to check the degree of slope the patio has before pouring concrete or laying patio bricks. This check is best done by using the inclinometer along a long, straight 2″ × 4″ board placed on the ground.

2. An inclinometer of this design can also be used to conduct studies of lake and ocean beaches. It may be used to measure the degree of beach incline. This might then be related to the type of vegetation and animal life found in each measured sector.

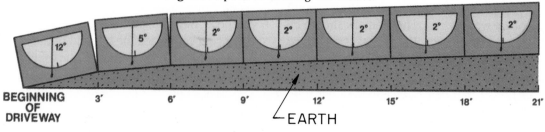

Fig. B. Steps in measuring the incline.

ACTIVITIES

Activity 2: Emphasizing Construction Safety

Objective

After completing this activity, you will know how to inform others of the safety hazards that exist in construction.

Materials Needed

- 100 sheets of plain white paper
- colored markers or crayons
- 12 scissors

Steps of Procedure

1. Your teacher will divide the class in half. One group will study safety practices as they apply to the construction of the substructure. The other group will study safety practices as they apply to the construction of the superstructure.

2. A symbol is a sign used to represent something else. Fig. A. Discuss the qualities a symbol must have if it is to be recognizable to all. Discuss the importance of color and design. You should remember the important points of this discussion when you begin to design your own safety sign.

3. Each student should draw two different and original safety signs. The signs should apply either to the substructure or the superstructure, depending on the group to which the student is assigned. Examples of these signs can be found at construction sites. Because each sign is the work of a different student, each sign will be different. Some of the signs may resemble in shape, color, and symbol — the signs actually used in the construction industries. Other signs will be different.

4. Tack up these signs around the classroom or lab area.

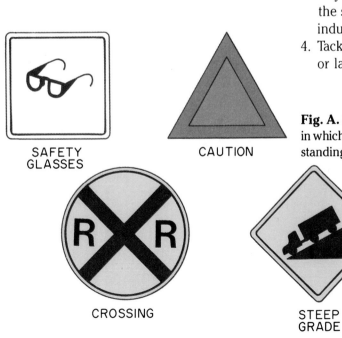

SAFETY GLASSES

CAUTION

CROSSING

STEEP GRADE

Fig. A. Some common safety signs. Note the way in which the symbol is used to promote quick understanding of the message of the sign.

SECTION

III

THE CONSTRUCTION COMPANY

Chapter 7: Organization of Construction Enterprises
Chapter 8: The Business of Construction

7 ORGANIZATION OF CONSTRUCTION ENTERPRISES

Terms to Know

board of directors

construction
management
companies

construction
superintendent

corporation

field engineer

field office

general contractor

home office

joint venture

partnership

project manager

proprietorship

specialty contractor

stock

stockholders

subcontractor

Objectives

When you have finished reading this chapter, you should be able to do the following:

- Define and explain the types of business ownership.
- Explain the difference between a general contractor and a construction management company.
- Describe the relationship between the general contractor and subcontractors.
- Explain the organization of a typical construction company.

Construction companies have many responsibilities in addition to the actual construction work. To carry out these responsibilities efficiently, a construction company must be well organized. First the type of company ownership must be decided and a chain of command must be established. If the company intends to specialize in a certain kind of construction, the specialization must be defined. Also, provisions must be made for hiring workers. The responsibilities of the workers must then be determined. As you can see, many factors affect the organization of a company. In this chapter, you will explore some methods of organization of construction companies.

The text mentions a few of the things that a construction company must consider when hiring workers. Can you think of some others?

TYPES OF OWNERSHIP

Ours is a *free enterprise system*. This means that businesses compete against one another for profit. Anyone is free to start their own business.

Construction is basically a business. The business is that of building structures. In this sense, a construction business is unique. However, a construction business has elements in common with other types of businesses. One such element is company ownership. There are three basic types of company ownership for any business: proprietorship, partnership, and corporation. Each of these has unique advantages and disadvantages.

Proprietorship

A company that is owned by one person is known as a **proprietorship**. This type of business can be called a one-person operation. Fig. 7-1. Usually the *proprietor*, or owner, uses his or her own money to get the company started. The proprietor makes all the decisions and has the sole responsibility for every part of the business. This includes buying materials, hiring workers, and seeing that the work is done. The proprietor is also responsible for all of the bookkeeping and accounting.

In a proprietorship, the owner keeps all the profit. However, if the company fails, the propri-

Fig. 7-1. The owner of a proprietorship makes all the final decisions. A proprietor assumes complete responsibility for all aspects of the operation of the company.

etor is personally responsible for its debts. If the company cannot pay its bills, the creditors (people to whom money is owed) can demand that assets, or property, be sold and the money given to them. If the company assets are not enough, the proprietor may have to sell personal assets to pay the creditors. That means the proprietor may have to sell a house, a car, or furniture.

Partnership

Two or more people can form a business known as a **partnership**. Fig. 7-2. Sometimes a partnership is formed when a proprietorship expands or when a parent brings a son or daughter into the business. Other partnerships may be formed by people who have similar interests in a certain type of business.

Most partnerships have a legal contract to protect the partners. This contract states the rights and responsibilities of each partner. Partners may divide the responsibilities in any way they wish. For example, one partner may supply the capital. Another may be responsible for the daily operation of the company. The partners can divide profits and responsibility for financial losses in any way they wish. If a formal agreement is not specified, all partners are assumed to be equally responsible.

Partnerships have several advantages over proprietorships. They offer additional financial resources and a wider range of skills. They retain the simplicity of a proprietorship. The financial responsibility in a partnership is not as great as that of a proprietorship. This is because more than one person is responsible for the company debt. However, in a partnership, authority may be divided. When the partners disagree on a major decision, they must be able to work with each other to resolve the problem.

Corporation

A **corporation** is a company that is owned by many people. Fig. 7-3. These people buy ownership of the company by purchasing shares of ownership called **stock**. The stock is sold in the form of *stock certificates*. Fig. 7-4. The more stock an individual owns, the more ownership

Fig. 7-2. These two people have formed a partnership. Together they own and operate their company.

Fig. 7-3. A corporation is owned by many people. This annual stockholders' meeting gives the owners an opportunity to voice their opinions.

he or she has in the company. The individuals who own the stock are called **stockholders**.

Even though a corporation is owned by many people, it is legally recognized as an individual. The law treats a corporation just as it would a person. A corporation can apply for loans, sign contracts, and sell merchandise just like a person.

The people who decide to start a corporation are called *incorporators*. Fig. 7-5. To start a corporation, the incorporators send an application to their state government for approval. This special application is called *articles of incorporation*. It lists the names and addresses of the incorporators. It also describes the proposed business, its location, and the number of shares of stock to be sold. After the articles of incorporation are approved, stock can be sold. A **board of directors** is elected by the stockholders to run the company.

Fig. 7-4. A stock certificate represents shares of ownership in a corporation. (*Goes Lithographing Co.*)

Fig. 7-5. These incorporators are discussing how to organize their new corporation.

The board of directors sets goals and makes policy decisions. Fig. 7-6. It also hires managers to take care of the daily business of the company.

Fig. 7-6. The board of directors meets periodically to set goals and company policy.

Did You Know?

The incorporation of businesses began in England in the early 1600s. At that time, businesses began to build up cash reserves. Looking for a place in which to invest it, some of them decided on the New World. Many of the companies that helped settle North America were, in fact, corporations. One example was William Penn's "Free Society of Traders," which helped settle what is now Pennsylvania.

A corporation has several advantages over either a proprietorship or a partnership. One advantage is that the owners have greater financial protection. If the corporation cannot pay its bills, it may declare *bankruptcy*. In bankruptcy, the company assets are sold to pay the debts. The stockholders lose the money they have invested. However, they do not lose their personal assets, such as homes and cars.

Another advantage of a corporation is that money can be raised at almost any time by selling additional shares of stock. The money from the sale of the stock is usually used for assets such as new equipment or an office building. The company thus becomes bigger and stronger.

A third advantage of a corporation is that it is legally recognized as an individual. Thus, it can exist beyond the lifetime of any single owner. The ownership of a corporation can be transferred easily without disrupting the business.

Corporations also have some disadvantages. For example, a corporation can only engage in business that is either stated or implied in its articles of incorporation. For example, a corporation that sells typewriters cannot suddenly decide to manufacture typewriters unless manufacturing is included in the articles of incorporation. Instead, the corporation would have to amend the articles of incorporation to include manufacturing before it could legally manufacture typewriters.

One of the biggest disadvantages of a corporation is the heavy taxes that are levied against it. A corporation must pay federal income tax. Then, after dividends are paid to stockholders, the stockholders must pay income tax on them. In effect, the profit of a corporation is taxed twice. In addition to income tax, corporations must also pay a franchise tax in some states. A *franchise tax* is a tax levied by a state for the privilege of operating as a corporation in that state.

Joint Venture

In the construction business, two companies sometimes combine into one for the purpose of working on a certain project. This type of ownership is called a **joint venture**. Fig. 7-7. For a specific period of time and for a specific purpose, the two companies agree to be known as a single company.

A joint venture benefits both companies. It gives them business that neither company alone would have been able to handle. By combining the forces of the two smaller companies, the larger company is able to handle a larger project. Both companies sign a contract that specifies all the details of the business. The contract defines the responsibilities of each company and the percentage of profits that each company will receive.

Did You Know?

The Chunnel is a 31-mile tunnel being constructed beneath the English Channel. The Chunnel will connect Great Britain and France. The Chunnel is being built by two companies. One company is an English company. The other company is a French company. To build the Chunnel, these two companies have formed a single company. The total cost of the Chunnel is huge—nearly $6 billion. There also are a number of construction engineering problems. In fact, the Chunnel is the largest civil engineering project in the history of Western Europe. Because the project is so large, it is more easily handled by two companies working closely together as a single unit.

Fig. 7-7. The Chunnel is an enormous construction project. It is a tunnel beneath the English Channel. It will link England with France. This massive project is a joint venture of two companies, one English and the other French.

For Discussion

This section has discussed the three basic types of company ownership: proprietorship, partnership, and corporation. If you were setting up a construction business, which type of company ownership would you prefer? Give reasons for your choice.

TYPES OF CONSTRUCTION COMPANIES

Most construction companies specialize in one type of construction, such as the building of highways. Other companies are very specialized. They do only one type of work, such as plumbing or landscaping. Still other companies specialize in management of construction projects.

General Contractors

A **general contractor** is a company that undertakes an entire construction project. Fig. 7-8. For example, a general contractor may undertake to build a building, a bridge, a power plant, or a highway. Whatever the project, the general contractor is responsible for all of the work that is done from start to finish.

A general contractor usually has an established crew of construction workers. In addition to these workers, most general contractors hire subcontractors for specific jobs. Together, the general contractor and the subcontractors are able to complete the project.

Specialty Contractors

Some construction work is done by companies that specialize in one type of construction job. They do only one kind of work, such as painting, electrical wiring, or paving. Such construction companies are called **specialty contractors**. Fig. 7-9. Specialty contractors provide a wide range of construction services to individuals as well as general contractors. For example, if you needed a new roof put on your house, you might call a roofing contractor.

When a general contractor hires a specialty contractor, the specialty contractor is known as a **subcontractor**. The general contractor sub-

Fig. 7-8. The construction project in the background is being built by a general contractor.

Fig. 7-9. Hanging wallpaper is a specialized type of work. It is usually subcontracted to a company that specializes in hanging wallpaper.

contracts, or hires out, a specific construction job to the subcontractor. A general contractor might typically hire subcontractors to do electrical wiring and painting. Other subcontractors might be hired to install plumbing systems, piping, and heating, ventilating, and air conditioning (HVAC) systems. Fig. 7-10. The landscaping work is usually done by a subcontractor that specializes in landscaping. Carpeting and other flooring materials are usually installed by subcontractors that specialize in flooring. Even elevators are installed by specialty contractors.

Although the general contractor hires the subcontractor to do a specific task, both companies are actually working for the owner of the property. The owner relies on the general contractor to act as the agent to hire the specialty contractors. The general contractor makes sure that the subcontractors do their work correctly and on schedule. The owner holds the general contractor responsible for the finished project. It is therefore to the contractor's advantage to see that the subcontractors do their work well.

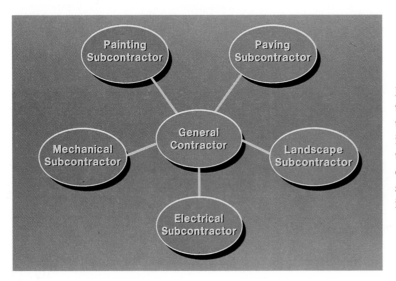

Fig. 7-10. The general contractor usually relies on subcontractors to do specialized work. These include the painting subcontractor, mechanical subcontractor, electrical subcontractor, landscaping subcontractor, and paving subcontractor.

Did You Know?

A canal is a specialized construction project. The Suez Canal provides a short route between the Mediterranean Sea and the Indian Ocean. The person mainly responsible for its construction was Ferdinand de Lesseps, a French diplomat. In building the canal, de Lesseps overcame a number of difficulties. After his success in building the Suez Canal, de Lesseps attempted to build a canal across the Isthmus of Panama. Here, however, he met with difficulties. Construction was halted. The canal was finally finished in 1914 by the United States.

For Discussion

What are the advantages in a general contractor hiring a specialty contractor to perform certain work on the project?

COMPANY ORGANIZATION

Every business must have some type of organization if it is to function efficiently. Construction businesses undertake two completely

Construction Management Companies

Construction management companies are those that manage construction jobs without doing any of the physical construction. A construction management company hires separate contractors to do each part of the work. Then it schedules and coordinates the work among the various contractors. Fig. 7-11. The construction management company is also responsible for overseeing the construction. It checks on the progress and makes sure the work is done properly.

Fig. 7-11. The construction manager coordinates all the work being done by the various contractors.

different kinds of work: office work and physical construction work. Most construction companies have found it convenient to maintain a permanent business office as well as an office at the site of construction. Therefore, construction companies are usually organized around two distinct offices: the home office and the field office. All of the organizational and business work is done at one of these two offices.

The Home Office

The **home office** is the company headquarters. Fig. 7-12. Most of the general work that is not actually physical construction is done at the home office. The home office is usually organized into three divisions: business administration, engineering, and construction.

Business Administration

The business administration division is responsible for business and financial operations. Fig. 7-13. Administrators are responsible for marketing, deciding which jobs to accept, and signing contracts and other legal documents. They also deal with labor relations and other personnel matters. In personnel management, the administration must be aware of any problems caused by job-related stress. They should be alert to ways of minimizing such stress. In addition, the business administration division is usually responsible for payroll preparation and accounting.

Engineering

The engineering division of the company is concerned with designing the structure of a

Fig. 7-12. The company headquarters is located at the home office.

Fig. 7-13. Business administrators keep the company running smoothly.

project on a given site. The **field engineer** oversees the project and makes certain that the building is laid out properly. Engineers also design the electrical and mechanical systems for a project according to the specific requirements of the owner. The engineering staff orders and supervises field surveys and helps develop the plans for the project. Estimators in the engineering offices estimate how long the job should take to complete, as well as the total cost of the job. Fig. 7-14.

Construction

The construction division of the company supervises the projects being built. This division is responsible for keeping track of the materials used and the actual amount of time spent on a project. Such work is called *project accounting*. The yards and shops are under the control of the construction group. The *yards* are where the company equipment is stored, and the *shops* are where it is maintained. The construction division also includes the project manager. The **project manager** is the person in the home office who is directly responsible for a certain construction project. Fig. 7-15.

Fig. 7-14. Estimators, working in the home office, determine how much work must be done to complete a project. They estimate what the project will cost.

Fig. 7-15. At the home office, a project manager may keep track of several projects at the same time.

The Field Office

The **field office** is a temporary office that is established at the construction site. It is usually a small building or a trailer. This office serves as the center of operations at the site. A clerk or secretary at the field office takes care of timekeeping reports, payroll, and other general paperwork. If the building project is large, the field engineer also has a desk at the field office.

The field office also serves as the home base for the construction superintendent. The **construction superintendent** is the person in charge of all construction proceedings. He or she is responsible for seeing that the construction work gets done correctly and on schedule. To keep the construction on schedule, the superintendent coordinates the efforts of the workers at the job site. Each of the different groups of workers has its own supervisor. Fig. 7-16. Each supervisor receives instructions and schedule information directly from the construction superintendent.

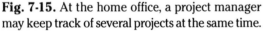

For Discussion

The home office is organized into three divisions: business administration, engineering, and construction. In order for the home office to operate efficiently, what qualities would be needed in the people working in the three divisions?

Fig. 7-16. The workers on a job can be more efficient if they are organized. The supervisor tells each worker what to do.

Construction Facts

CONSTRUCTION BOSS

Mary Smith, a 43-year-old single mother, has turned a borrowed $1,000 into a construction company that now does almost a million dollars' worth of business a year. Ms. Smith is the president of her own construction company. She is one of a small number of black women who head construction companies. How did she get into the construction business?

In 1987, she was laid off from her job as a waitress. Determined to provide for her family, she borrowed $1,000 from her uncle. With this money and the cooperation of a few other people, she incorporated her company.

At first she encountered resistance everywhere she turned. The Small Business Administration turned down her loan applications. People simply could not understand why a former waitress would want to operate a construction company. Her persistence, however, has paid off.

Now her company does both general contracting and subcontracting. The company has renovated several apartment buildings under contracts of up to $150,000. Her company has also worked as a painting subcontractor for other construction companies.

People thought that Mary Smith was crazy to get into the construction business. Nevertheless, through her determination and hard work she has built a company that now has 17 employees.

CHAPTER 7

R E V I E W

Chapter Summary

Construction is a business. Company ownership can be organized as a proprietorship, partnership, or corporation. There are advantages and disadvantages in each type of ownership. A joint venture is another type of ownership. A general contractor is a company that undertakes an entire construction project. A specialty contractor specializes in one type of construction job. Construction management companies manage construction projects without doing any of the work. The field engineer oversees the construction project. The construction superintendent is the person in charge of all construction proceedings.

Test Your Knowledge

1. In a proprietorship, how many owners does a company have?
2. Who is responsible for company debts if a partnership fails?
3. What is the name of the type of company that sells shares of ownership to many different people?
4. What is stock?
5. What is the name of the type of construction company that is formed from two smaller companies for the purpose of doing a specific project?
6. What is the difference between a general contractor and a subcontractor?
7. Which type of company is responsible for seeing that the project is completed properly but does not take part in the actual construction?
8. What are the two kinds of construction company offices?
9. Name two responsibilities of the field engineer.
10. What person is responsible for coordinating the work of the subcontractors and seeing that the work runs smoothly?

REVIEW

Activities

1. Look under the heading *Construction* in the Yellow Pages of your telephone book. From the company names that are listed, see if you can guess whether each one is a proprietorship, partnership, or corporation.

2. Most construction projects have a large sign posted near the front of the site. Visit a construction site. Locate the field office and look at the sign. Answer the following questions:

   a. Is a general contractor or a construction management company in charge of building the project?

   b. In what city is the home office of the general contractor or the construction management company?

8 THE BUSINESS OF CONSTRUCTION

Terms to Know

bar chart
bids
bond
construction
  superintendent
contract
cost accounting
cost-plus contract
critical path method
  (CPM) chart

incentive contract
lump-sum contract
negotiate
overhead
payment bond
performance bond
project accounting
project control
project manager
unit-price contract

Objectives

When you have finished reading this chapter, you should be able to do the following:

- Explain the difference between negotiating and bidding.
- Explain the four main steps in the bidding process.
- Describe four types of contracts.
- Explain how bonds protect the owner of a construction project.
- Describe two types of construction schedules.
- Describe project control procedures.

SUPER INSULATED HOME

sensible solar homes

- DOUBLE EXTERIOR WALL R-38
- ATTIC INSULATION R-44
- AIR TIGHT CONSTRUCTION
- INSULATED STEEL EXT. DOORS
- INSULATED BAND JOIST R-19
- SOFFIT/RIDGE VENTILATION
- 2" FOUNDATION INSULATION
- WINDOW GLASS LOW-E
- HIGH EFF. FURNACE/W. HEATER
- SETBACK THERMOSTAT

noah herman Sons / Builders 686-2230

Construction companies get most of their business in one of two ways. Sometimes the owner of a project approaches a construction company and asks for an estimate on the project. Fig. 8-1. The owner and the construction company **negotiate**, or discuss the terms of, a contract for the company to do the project. A contract is needed to spell out the responsibilities of both the builder and the owner. The cost of the project and the finish date are also negotiated. When both parties agree on the terms, the contract is signed and construction is started.

The other way for a construction company to get business is to compete with other construction companies for a particular job. In this case, each company **bids**, or quotes a price for which it will do the job. Construction companies obtain most of their work through this competitive bidding process.

THE BIDDING PROCESS

The bidding process is begun by the owner after the designing and engineering work on a project is finished. The owner sends out invitations to bid on the project. The bid invitation lets the construction companies know about the proposed project so they can decide whether they are interested. There are several ways of inviting companies to bid on a project. One way is to contact several companies individually and ask them to bid. Another popular way is to issue a public invitation. A series of advertisements is placed in newspapers and construction industry magazines. Fig. 8-2. Each advertisement states

- the type of work.
- the location of the project.
- where to get plans and specifications.
- the time and place of the bid opening.

Preparing a Bid

First a construction company must decide whether it has the staff, time, and resources to do the job. If so, office personnel prepare a bid. A key person in preparing the bid is the *estimator*. The estimator must be very careful when he or she prepares the bid. If the bid is too high, the company will not get the job. If the bid is too low, the company may get the job, but it will not make any profit on the project.

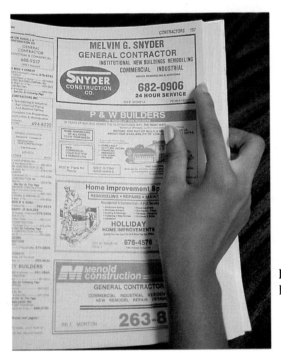

Fig. 8-1. Many contractors are listed in the Yellow Pages of the telephone book.

The Harbor Authority

Sealed proposals for the following contract will be received by the Chief Engineer, Room 32S Indiana Boulevard #1, Peoria, Massachusetts 10048, until 2:30 P.M. on the date indicated and will then be opened and read in Room N. 32E. Contract documents may be seen in Suite 5147 - 51st Floor and will be furnished upon request. Please call first for availability of Contracts. Questions by prospective bidders concerning any one of the contracts should be directed only to the person whose name and phone number is listed for the contract in question. No deposit is required.

Contract DAR-111-025 — Peoria International Airport — Terminal C — International Departures Facility - Bids Due Tuesday, December 11. Direct questions to Mr. Dennis Zempel.

The Harbor Authority

Fig. 8-2. The owner of a project may issue a public invitation to bid. This bid invitation was published in a construction magazine.

Estimating the Cost of Materials

The estimator studies the plans and specifications of the job carefully. Then he or she uses a special form to make a list of the materials needed. Fig. 8-3. The price of each material is found and multiplied by the amount of material needed. The resulting amounts are then added together to find the estimated cost of the materials.

Fig. 8-3. The estimator uses a form similar to this one to estimate the cost for the materials needed for each part of the project.

XYZ CONSTRUCTION COMPANY

ESTIMATE FOR _University Classroom Bldg._

ESTIMATOR _____ DATE _____
CHECKED BY _____ DATE _____

SEC/PROD	ITEM	QUANTITY	UNIT	MATERIAL	UNIT	EQUIP. OR SUB.	UNIT	LABOR	TOTAL
	Place wall footing	1825 cy	47^{10}	860			5^{60}	107	962
	Place foundation walls	7900 cy	47^{10}	3721			5^{60}	442	4163
	Place slab on grade	2650 cy	47^{10}	1248			4^{67}	124	1372
	Cure and protect	9500 SF	0^{15}	1425			0^{03}	224	1649
	Trowel finish	9500 SF					20	1900	1900
	Float finish	9500 SF					0^{18}	1730	1730
	Expansion joints	196 LF	0^{46}	90			0^{35}	69	159
	Concrete pump					2800	128	1280	4080
	4 Power trowels	4 Days			100^{50}	400		512	912
	Foreman	20 Days					128	2560	2560
				7344		3200		8943	19487
	Tax		5%	368					
	PT						23%	2057	2057
	FB						23%	2057	2057
									23969

_____ EXTENDED BY _____ DATE _____
_____ CHECKED BY _____ DATE _____

Estimating the Cost of Labor

The cost of labor is estimated next. To determine labor costs, the estimator looks in labor reference books to find the standard amount of time needed to do certain jobs. In figuring labor costs, the contractor needs to take into consideration the number of workers needed, how long they will be needed, and the wage rate at which they will work. With this information, the estimator can estimate the total cost for labor for the job.

Estimating the Cost of Equipment

Equipment expenses must be estimated, too. Fig. 8-4. There are two kinds of equipment expenses. One kind of expense is the price or rental cost of the machines and equipment. The other is the cost to run and maintain the machines and equipment.

Getting Estimates from Subcontractors

Some of the special construction work may need to be subcontracted. The subcontractors will also need a set of plans and specifications to determine how much their part of the project will cost. The subcontractors figure out their own costs and profit. Then each subcontractor tells the general contractor his or her price for doing the work. The contractor has to include those prices in the bid.

Estimating Overhead Expenses

Another factor that must be considered is the company's overhead. **Overhead** is the cost of doing business. Costs for electricity, water, telephone service, office salaries, and postage are examples of overhead costs. Other overhead costs include the costs of advertising, insurance, and office rent. The estimator figures the approximate overhead for the project and includes it in the estimate.

Estimating Profit

The last figure that the estimator must calculate is the company's profit. This figure is difficult to estimate. If the profit is too high, the bid will be too high. This will eliminate the company from the competition. If the profit is too low, the job may not be worthwhile.

Fig. 8-4. The estimate includes rental or purchase costs as well as operating costs of all the equipment needed for a project.

To put the bid together, the estimator adds the individual costs. He or she double-checks the figures for accuracy. A supervisor usually checks the figures also. Mathematical errors can cost the company a great amount of money. A large error might even bankrupt a company.

Analyzing the Bids

The bids from the contractors are submitted to the owner in sealed envelopes. At a specified time, the owner holds a *bid opening*. Each company that submits a bid sends a representative to the bid opening to hear all the bid prices. The sealed bids are opened one at a time. As each bid is opened, the bid price is read aloud so everyone can hear it. Fig. 8-5.

The contract is not usually awarded at the bid opening. The owner takes the time to analyze all the bids before he or she decides which construction company to hire. The analysis is done by the owner or by the architect or engineer for the project. All the bids are studied very carefully and are compared to one another. Those that do not meet the specifications are eliminated. The reputation of each of the remaining contractors is checked thoroughly to see that the company has been reliable on other jobs. Finally the owner decides which company will get the work. If more than one company qualifies, the one with the lowest bid is usually chosen.

Fig. 8-5. Sealed bids are opened and read aloud at a bid opening.

For Discussion

By opening a job to bids by various companies, the owner helps to make sure that the project is built at the lowest possible price. Can you think of any circumstances in which the owner might not award the contract to the company with the lowest bid?

CONTRACTS AND LEGAL RESPONSIBILITIES

Once a builder has been selected, either by negotiation or by competitive bidding, a contract is prepared. The **contract** is a written agreement between the owner and the contractor. The owner and the contractor are known as the *parties* to the contract. The responsibilities and rights of each party are stated in the contract. The contract contains information about the amount of work to be done, the price to be paid, and the method of payment. Both parties sign the contracts. Fig. 8-6. The contract is a legal document. If one of the parties does not keep his or her part of the contract, he or she can be taken to court. The dispute will be settled there.

Did You Know?

Contracts arose because people in business needed a way to make sure that a promise could be enforced by law. The Romans had developed a system of law dealing with contracts. During the Dark Ages, from the fall of Rome (476 A.D.) to about 1000 A.D., society became more agricultural and town life decreased. At this time, contracts became less important. Around 1100 A.D., however, economic life began to flourish again. Then contract law became more important.

Fig. 8-6. The contract is signed by representatives of the owner and contractor.

Kinds of Contracts

There are four basic kinds of contracts. Each one is suitable for a particular type of job. The owner and the contractor must choose the right contract for each job.

Lump-Sum Contracts

A **lump-sum contract** is sometimes called a *fixed-price contract*. It is, as the name implies, a contract in which a lump sum (fixed price) is paid for the work to be done. The fixed price is agreed upon before the work begins. The sum may be paid in several payments. The final payment is made when the work is completed satisfactorily.

The advantage of the lump-sum contract is that all the parties know what to expect. The owner knows how much the project will cost. The construction company knows how much money it will receive.

Cost-Plus Contracts

In a **cost-plus contract**, the owner agrees to pay all the costs of construction, including materials and labor. In addition, the owner agrees to pay the contractor an extra amount to cover the contractor's overhead and profit. There are two types of cost-plus contracts. One is cost plus a fixed fee. The other is cost plus a percentage of the cost.

The *cost-plus-fixed-fee* contract states that the owner will reimburse, or pay back, the contractor for all the expenses of construction and will add a specific amount of money to that. The extra amount is agreed upon before the contract is signed. For example, suppose you agree to paint a neighbor's garage. Your neighbor promises to reimburse you for the paint and to pay you $20 extra. You know before you begin that your profit will be $20.

The *cost-plus-percent* contract is similar to the cost-plus-fixed-fee contract. The difference is that the extra amount is a percentage of the total cost

rather than a fixed amount. Using the same garage-painting example, suppose your neighbor agrees to reimburse you for the cost of the paint and pay you an extra 20 percent. If the paint cost $105, your neighbor will owe you $105 for the paint, plus 20 percent of $105, or $21. Thus, your neighbor will owe you $126. How much would the neighbor owe if the paint cost $85.

Incentive Contracts

An **incentive contract** is designed to reward or penalize the contractor, depending on when the job is completed. If the job is finished before the agreed-upon date, the contractor is rewarded with an amount of money that is specified in the contract. Generally, under this type of contract, the contractor gets to keep part of the savings if the company can keep costs lower than the original estimate. If the job is not done by the specified date, the contractor is penalized a certain amount of money.

Unit-Price Contracts

Sometimes it is difficult to estimate the amount of work that needs to be done on a project. A **unit-price contract** is good for that kind of job. The contractor gives the owner a unit price. The contractor will charge by the unit of work. Consider the garage-painting example again. Suppose you agree to paint the garage for $.50 per square foot. The unit is 1 square foot and the price per unit is $.50. Thus for every square foot you paint, your neighbor will pay you $.50. If you paint one 8-foot-by-20-foot wall, the neighbor will owe you $80 (8 ft. $\times$ 20 ft. = 160 sq. ft.; 160 sq. ft. $\times$ $.50 = $80). This kind of contract is often used for paving and highway repair work.

Bonds

The owner of a project has the right to expect the construction company to do the job well, complete the project on time, and pay its workers. The owner depends on bonds to help make sure all these requirements are met. A **bond** is similar to an insurance policy. Bonds are meant to provide protection for the owner in the event the contractor does not follow the terms of the contract. To get a bond, a construction company must have a good reputation and financial dependability. The contractor pays a fee to the bonding company. In return for the fee, the bonding company issues the bond. Depending on the contractor's needs, the bond may be either a performance bond or a payment bond.

Performance Bonds

A **performance bond** guarantees that the contractor will build the project according to the agreement. If for some reason the contractor is unable to finish the job, the bonding company is responsible for seeing that the rest of the work is done. This kind of bond guarantees that the owner will not have to pay additional money to another contractor to have the job completed.

It is very unusual for a bonding company to have to take over a project to finish it. Bonding companies are careful about the contractors to whom they give bonds. Since most contractors want to stay in business, they work to protect their reputation.

Payment Bonds

A **payment bond** is a guarantee that the contractor will pay his employees, subcontractors, and suppliers. This kind of bond is important because if someone is not paid, he or she can file a legal claim against the owner. This means that the unpaid party can claim ownership of the new structure. A payment bond prevents situations such as this from happening. If the contractor fails to pay a subcontractor, for example, the bonding company makes the payment. Thus, the owner is protected.

Safety and Liability

The construction company must also meet certain other legal obligations. The contractor must keep the site as safe as possible for the workers. The company may also be liable for any accident that happens on the site. To be liable is to be legally responsible. Most construction companies are therefore very concerned about safety and liability. The safety and protection of their workers, the general public, the site, and the property close to the site are important considerations.

HEALTH & SAFETY

For someone not familiar with them, construction sites can be dangerous places. There may be deep holes for building foundations. There may be a variety of building materials stored there. Accidents are possible. The workers on construction sites should be properly clothed. They should also be familiar with the safety practices that should be followed on a construction site. The contractor does not like to have others on the site. For this reason, most large construction sites are fenced in. Nearby walkways may be covered. Don't trespass on a construction site.

Protecting the Workers

Construction work can be dangerous. Precautions must be taken to avoid as many accidents as possible. Workers in certain areas, called *hard hat areas*, wear hard hats to protect them from falling objects. Fig. 8-7. Other examples of personal protection include safety glasses and

Fig. 8-7. Hard hats help protect workers from falling objects.

gloves, which must be worn for certain operations. Safety glasses prevent flying objects such as slivers of metal or wood from getting into the workers' eyes. Gloves protect the workers' skin from exposure to harmful or irritating chemicals.

Keeping tools and equipment in good condition is another way to prevent accidents. Ladders must be sturdy and in good condition. The railings that are provided around scaffolds, or high platforms, to keep workers from falling must be solid. Fig. 8-8. It is common for a construction company to have weekly safety meetings to help the workers be more alert and aware of safety. Inspections may also be held periodically to identify unsafe conditions. Once they are identified, the problem areas should be corrected immediately.

Fig. 8-8. Railings are used to prevent workers from falling when they are working above the ground.

The Occupational Safety and Health Administration (OSHA) is an agency of the federal government. It is responsible for making sure that workers have a safe place to work. OSHA inspectors can visit a job site and perform a safety inspection. Fig. 8-9. If there is an unsafe condition, the inspector can shut down the job until the problem is corrected.

The government also requires that a company carry a special kind of insurance called *workman's compensation*. This insurance pays the medical expenses of a worker who is injured on the job.

Protecting the Public

The contractor is also responsible for making sure the public is not endangered by the construction. For example, the construction company may build a fence around the site to prevent people from coming too close. Fig. 8-10. If work is going on overhead, a special tunnel or canopy is built to prevent things from falling on passersby. Barricades may be used to block off a road or to redirect traffic. Sometimes a flagperson may be used to direct traffic around a dangerous condition. Fig. 8-11.

Fig. 8-9. OSHA inspectors check the site to make sure that safety laws are being obeyed.

Fig. 8-10. A fence around the job site keeps unauthorized persons out of the site and away from danger.

Fig. 8-11. A flagperson controls the traffic on a road construction project.

For Discussion

You have probably seen construction sites in your city or town. On these sites, what measures have the contractors taken to protect the public?

Protecting the Surroundings

The construction company must also provide protection for the area near the job site. Many times there are trees or other buildings near the site of the proposed structure. These must not be damaged while the new structure is being built. If the property around the site is damaged, the contractor is responsible for making the necessary repairs.

ORGANIZING THE JOB

Once a construction company gets a contract for a job, it must begin organizing for that job. A **project manager** is appointed to coordinate the money, workers, equipment, and materials for the job. He or she develops a schedule and plans for the storage of goods at the job site. A method of keeping track of all construction activities must be devised. Without organization, the work cannot be done efficiently.

Scheduling

The schedule of work tells what tasks must be done and when they must be completed. Two methods of scheduling are commonly used by contractors. With either method, the project manager can estimate the amount of time it will take to do each part of the job.

Bar Charts

The **bar chart** is easy to read and interpret. Figure 8-12 shows an example of a bar chart. The months are listed across the top of the chart. All the major jobs are listed down the side of the chart. A bar is used to show the starting and completion dates for each job.

Critical Path Method

The **critical path method (CPM) chart** is a diagram made of circles and lines. Fig. 8-13.

Fig. 8-12. A bar chart is used to show a construction schedule.

Project: _____

Date: _____ By: _____

OWEN CONSTRUCTION
CONSTRUCTION SCHEDULE

	July	August	September	October	November	December	January	February	March	April
Excavation and grading										
Foundations - formwork - walls										
Rebar										
Concrete										
Suspended slabs										
Slab on grade										
Plumbing underground										
Plumbing above ground										
Electrical underground										
Electrical above ground										
Mechanical - air conditioning										
Roofing										
Lath and plaster										
Millwork - doors - windows										
Painting										
Paving and landscaping - parking										
Hardware										
Final inspection - pick up comp.										

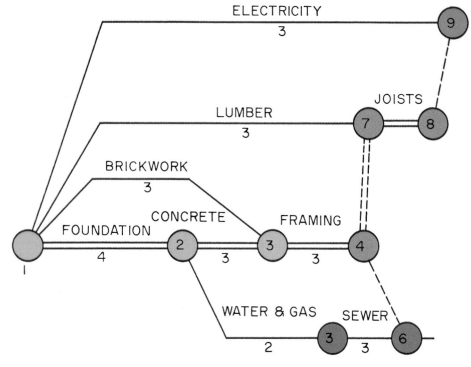

Fig. 8-13. A critical path method (CPM) chart.

Each line and circle has a meaning. This kind of schedule is useful because it shows the critical parts of the job clearly. Certain parts of a project have to be done before other parts. For example, the concrete cannot be poured until the reinforcing steel has been set in place. In turn, the steel cannot be placed until the concrete forms are built. Therefore, building the forms is a critical task.

Organizing the Job Site

The construction company must also have a plan for doing the work efficiently. One of the first steps in organizing the work site is to gain access to it. Usually streets already provide access to the site, but sometimes a *haul road* must be built to reach the site. Fig. 8-14.

Once access is established, several kinds of temporary structures are set up at the site. In many cases, the construction company brings in a trailer to be used as an office. Fig. 8-15. This will be the field office headquarters for the company. Portable restrooms are brought in. Sheds are built or trailers are brought in for storing materials. Some materials, such as insulation, must be protected from the weather. Other materials, such as pipe valves, are expensive and must be protected from damage and theft.

Utilities are needed temporarily at the job site during construction. Electricity is needed for power saws and other power tools. Telephones are necessary at the field office. Water is needed for drinking and washing. A waste container is needed for waste and scrap materials. Fig. 8-16.

Fig. 8-14. This haul road provides access to the construction site. It will be removed after the job is finished.

Fig. 8-15. A trailer is usually brought to the site to serve as an office.

Fig. 8-16. Even the location of waste disposal containers must be planned.

Ordering Materials

If materials are not at the site when they are needed, the job will be delayed. Ordering the materials at the right time is therefore an important job. The construction company's purchasing agent orders the materials according to the estimate that was prepared for the bid. He or she works to get the best price and delivery on the correct materials. The purchasing agent must be familiar with the delivery times for different materials. It may take a long time for some materials to be delivered. Other materials can be brought quickly to the site. Materials are ordered according to the estimate that was prepared for the bid.

For Discussion

You may have noticed certain temporary buildings at construction sites in your town or city. From what you have read in the text, can you identify the uses for these temporary buildings?

Fig. 8-17. A construction superintendent gives directions to the construction workers.

CONTROLLING THE JOB

Construction must be carefully controlled if the project is to be done well and on schedule. No matter how well a project has been planned, problems can arise. The person who controls all activity at the site is the **construction superintendent**. He or she must be aware of any problems and make corrections when they are needed. At the same time, he or she must keep a close watch on the materials that are bought and how much they cost. The cost of materials is checked through an accounting system.

Did You Know?

For centuries, accounting entries were made using pen and ink. Now, several accounting programs are available for computers. However, the computer has not changed the basic rules of accounting. Accounting rules remain the same. Computer accounting programs have, though, made accounting easier. They have made it easier especially for those who are not accountants. Computer accounting programs allow many small businesses to keep careful track of income and expenses. They could, of course, have done this without using a computer accounting program. In some cases, however, it might not have been as easy.

Project Control

Project control is the process of giving directions and making sure the job is done properly and on time. The construction superintendent is responsible for directing and monitoring the workers. He or she is almost always on the site to perform these functions. Fig. 8-17.

Another aspect of project control is monitoring the work. The construction superintendent sees that the working drawings are followed and that the correct materials are being used. He or she also checks the progress of each step to keep the project on schedule.

Project Accounting

The contractor must account for the progress made at the job site. To keep track of the progress, the contractor must keep accurate records of what has been done. This is known as **project accounting**. The contractor keeps track of how many carpenters are working, how much concrete has been used, and how close to schedule the project is progressing. Daily reports are turned in by the foreman for each type of worker, or *trade*. For example, the foreman for the bricklayers is responsible for a daily report of the number of workers and how much work is accomplished. Fig. 8-18. Weekly reports are sent to the home office. From these reports the master schedule is updated. The date on which the work is completed is checked against the schedule. The contractor meets periodically with the owner, architect, and engineer to report on the progress of the project. Such meetings are scheduled weekly or monthly, but they may be held whenever they are needed. Fig. 8-19.

Cost Accounting

When the construction company made its bid, it had to make an estimate of what it would cost to build the project. If the company uses more than the estimated amount of materials the costs will rise and the company will lose money. **Cost accounting** is the procedure by which the company keeps track of the costs of the project. The people who do the accounting for the company are very thorough. They check carefully all the

CARLSON CONSTRUCTION COMPANY, INC.
PRODUCTION SUMMARY REPORT

Job Name: Westside Hospital   Job No. 84-163

Foreman's Name: Fred Smith   Clock No. 1508

Date: May 1

Breakdown Number

	Mon.	Tues.	Wed.	Thurs.	Fri.	Other	TOTAL
455.01 Pipe	102'	56'	101'	75'			334'
455.03 Pipe	30'		54'				84'
455.05 Elbow	6		8	2			16
455.06 Tee	1	1	1				3
455.09 Valve		1		2			3

Fig. 8-18. This report was submitted by the plumbing foreman. It gives a summary of the materials used on the project on May 1.

bills for materials and labor. If more money is being spent than was estimated, they notify the company managers so that corrections can be made.

The overall progress chart helps the contractor keep track of the job costs. As the work progresses, the appropriate changes are made in the cost of the project. This chart helps control the total cost of the project. Keeping good records of the costs also helps the company on its next bid. It can bid more accurately the next time because the estimator knows how much time and material it took to do the last job.

Fig. 8-19. At a progress meeting, the architect or engineer and the contractor discuss the progress of the job with the owner.

Did You Know?

Bookkeeping is the record-keeping part of accounting. Bookkeeping has been practiced for hundreds of years. In the 1400s, the Italians developed several manuals for bookkeeping. In fact, the modern system of bookkeeping developed in Italy in the 1200s. There still exist a set of bookkeeping records from the year 1340. All of the entries in the records were made correctly.

For Discussion

In this section, you have read about several methods used to keep track of the progress and cost of a project. What advantages are there in being able to keep careful track of the work being done on a project and the money being spent?

Construction Facts

SUPER SCHEDULE SPEEDS STRUCTURE

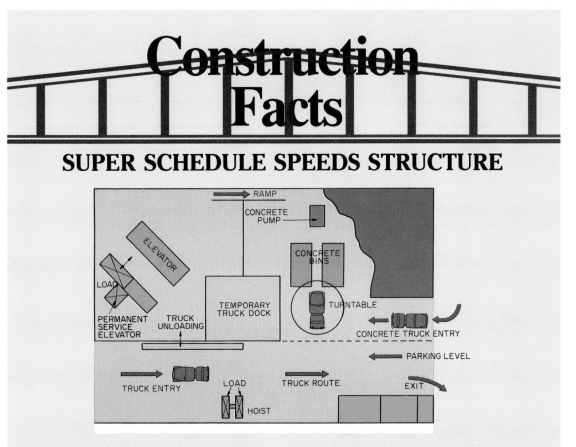

Careful planning by a Canadian firm, Olympia and York, helped speed up the construction of an 11-story Los Angeles high-rise building. The company developed a "super schedule"—a master plan for the construction site. Every detail of construction was carefully planned and scheduled. The plan included many advanced techniques, all planned to work together to save time and money. Some of the timesavers included using more and larger cranes and elevators, fine-tuning delivery schedules, controlling movement of all materials from a "command center," and systematizing storage.

Materials were delivered daily as they were needed. Few materials needed to be stored on the site. Companies delivering the mate-rials had to sign up to reserve unloading positions and hoisting times. Each type of material, including drywall and glass windows, was stored in the same designated location on each floor. This reduced handling, congestion, and confusion.

One of the unique ideas was the use of a one-way delivery lane with a rotating turntable. Trucks drove onto the turntable, which was rotated half a turn. Then the truck was unloaded. While that was happening, another truck was driving onto the turntable. This allowed deliveries to be made twice as fast. This super schedule and advanced planning eliminated 152,000 hours of work time. It also shaved $3 million off the $96 million job.

CHAPTER **8**

R E V I E W

Chapter Summary

Construction companies get their business by giving an estimate on a project or by bidding on a project. A bid must be carefully prepared. The cost of materials, labor, equipment, and overhead must be considered, among other expenses. A contract is prepared between the builder and the person hiring the builder. This contract outlines the rights and responsibilities of each party. There are lump-sum contracts, cost-plus contracts, incentive contracts, and unit-price contracts. A bond might also be required. This protects the owner if the builder does not follow the terms of the contract. There are performance bonds and payment bonds. The construction company is obligated to protect its workers and the public. A construction job is organized to coordinate money, workers, materials, and equipment. In controlling the construction project, the construction superintendent uses the practices of project control, cost accounting, and project accounting.

Test Your Knowledge

1. What are two ways in which a construction company can get business?
2. Name four types of information that should be included in a bid invitation.
3. Who is the key person in the bid preparation process?
4. Name five kinds of costs that an estimator must consider when preparing a bid.
5. In which type of contract is the total price agreed upon before construction begins?
6. What is the "'incentive'' in an incentive contract?
7. How does a payment bond protect the owner of a project?
8. What are two safety precautions that construction workers can take?
9. What are the names of the two common methods of scheduling?
10. Who is responsible for controlling all the work at the site?

REVIEW

Activities

1. Look in the classified advertising section of your local newspaper. Find an advertisement that is an invitation to bid. Clip the bid invitation from the newspaper and attach it to a sheet of notebook paper. Beneath it, state in your own words what the invitation to bid is about.

2. Visit a construction site. Make a list of the safety and protective devices you see being used. Tell who the device is protecting (workers, the public, etc.).

3. Make a bar chart schedule of your daily activities. Across the top of the chart, list the hours of the day (7 A.M., 8 A.M., etc.). Down the left side, list the tasks you normally do (eat breakfast, ride the bus, etc.). Estimate the beginning and ending time for each task. Then draw the bar chart. Use the chart to check on your actual time use for one day.

Activity 1: Estimating Wall Construction Materials

Objective

After completing this activity, you will know how to calculate needed amounts of insulating materials.

Materials Needed

• Examples of various insulating materials. These examples include batts, blankets, reflective materials, a loose fill, and rigid materials.

Steps of Procedure

1. Measure the perimeter of the structure in Fig. A. The perimeter is the outside boundary of the structure. Also measure the ceiling height. Multiply the perimeter by the ceiling height.
2. Measure the area of the doors and windows. Multiply the width by the height.
3. Deduct the area for doors and windows. Many carpenters will deduct only the area of large windows and window-walls. They will disregard the area of doors and smaller openings. This allowance will make up for loss in cutting and fitting. It also will provide for additional material needed around plumbing pipes.
4. For an example, refer to Fig. A.
   Perimeter = $30' + 40' + 36' + 20' + 6' + 20'$
   Perimeter = $152'$
   Ceiling height = $8'$
   Area: perimeter × height = $152' \times 8'$

Window wall = $12' \times 8'$
Window wall = 96 sq. ft.
Net area = $1216 - 96$
Net area = 1120 sq. ft.

5. Fill insulation comes in bags that usually contain 3 or 4 cu. ft. The number of cubic feet of fill insulation required can be calculated as follows:
   Area = 1200 sq. ft.
   Thickness = 4 in. = $\frac{1}{3}$ ft.
   Cu. ft. required = $1200 \times \frac{1}{3} = 400$
   Less 10% (allowance for joists 16" on center = $400 - 40$)
   Net amount = 360
   Number of bags (4 cu. ft. = 90 bags)

Fig. A. The perimeter dimensions of a house.

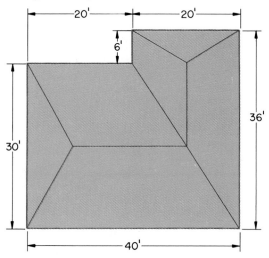

Activity 2: Conducting Soil Fracturing Tests

Objective

After completing this activity, you will be familiar with the concept of soil fracturing or shearing. Soil shearing is an important soil characteristic that must be tested before construction begins. By constructing the following apparatus, you will see the shearing characteristics of different soil samples. Testing soil by use of a soil fracturing test is one of many responsibilities of the engineers in charge of site preparation.

Materials Needed

- 1-4' length of 2 × 4
- 2-8" × 12" × ½" plywood sheets
- 32-1¼" × #8 flathead wood screws
- 4-4" × 8" × ½" plywood sheets
- 1-¾" eyelet
- 1 pulley (½" shaft)
- 1-3" × ½" metal rod (pulley shaft)
- work station with 2 vises
- drill and ⅛" drill bit
- 10' of string
- 25 lbs. of weight
- 1 stopwatch
- 2-12" metal rules
- 1 roll of tape

Steps of Procedure

1. Cut the 2 × 4 into four 12" lengths.
2. Fasten the 12" pieces of 2 × 4 to the outer edges of one of the sheets of 8" × 12" × ½" plywood. See Fig. A.

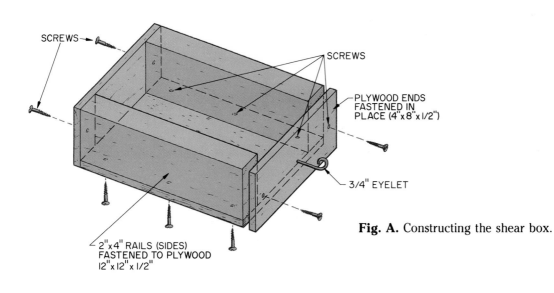

SCREWS

SCREWS

PLYWOOD ENDS FASTENED IN PLACE (4"x 8"x 1/2")

3/4" EYELET

2"x4" RAILS (SIDES) FASTENED TO PLYWOOD 12"x 12"x 1/2"

Fig. A. Constructing the shear box.

ACTIVITIES

3. Repeat Step 2, with the second 8″ × 12″ × ½″ plywood.
4. Fasten the 4″ × 8″ × ½″ plywood sheets to the ends of each half of the shear box. See Fig. A.
5. Fasten the ¾″ eyelet to one-half of the shear box at the center of one end. See Fig. A.
6. Drill six moisture escape holes in the top and bottom plywood pieces at various locations.
7. Place the bottom half of the shear box in a vise. (The bottom half is the part without the eyelet.) Clamp the box between the 12″ lengths.
8. Place the top half of the shear box on top of the bottom. The top half should slide easily on the bottom half. (The eyelet should be facing the second vise station.) Fig. B.
9. Tape each metal rule to the 12″ side of the shear box. The edges of the rules should meet at the split line of the shear box.
10. Place the pulley on the metal shaft and fasten into the second vise station.
11. Fasten the string to the eyelet around the pulley. Let the free end hang toward the floor.
12. Remove the plywood top from the top half of the shear box. Fill the box with a soil sample. Pack the soil firmly into place.
13. Replace the plywood top.
14. Attach the weight to the free end of the string. Place a box beneath the weight. Start the stopwatch.

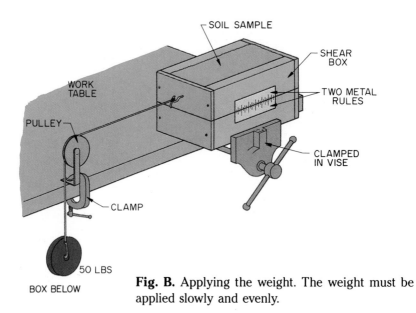

SOIL SAMPLE
SHEAR BOX
WORK TABLE
TWO METAL RULES
PULLEY
CLAMPED IN VISE
CLAMP
50 LBS
BOX BELOW

Fig. B. Applying the weight. The weight must be applied slowly and evenly.

BOX BELOW

15. Record the distance the top half moves in each minute. Continue to record these distances each minute until the stopwatch has run for ten minutes, or until a complete shear occurs. See Fig. C.
16. Now that you have collected all your data, construct a graph as shown in Fig. D.
17. Plot the shear value for each minute of testing.
18. Connect each of the plotted points in successive order to make a line graph.

SHEAR VALUE (in inches)	MINUTES
0.25	1
0.75	2
1.25	3
2.00	4
COMPLETE SHEAR	5
-	6
-	7
-	8
-	9
-	10

Fig. C. Recording the shear measurements.

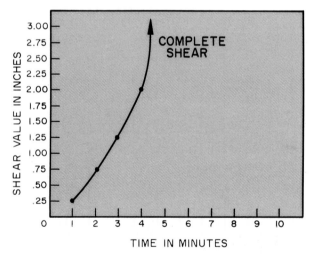

Fig. D. Plotting the shear values on a graph.

ACTIVITIES

Activity 3: Understanding the Systems Approach to Construction

Objective

After completing this activity, you will be able to identify the main parts of the construction industry when that industry is viewed as a system.

Materials Needed

- Paper (two 8½″ × 11″ sheets per team)
- Pencil (one per team)
- Scissors (optional, one pair per team)

Steps of Procedure

1. Your teacher will organize the class into teams. There will be two persons to a team. He or she will then familiarize you with the following terms: system, input, resource, process, output, and feedback. Definitions for each of these terms follow.

   System: A whole made up of several different parts.

   Input: Something, such as information or energy, that is added to a system.

   Resource: An object or action that can be used to solve a problem.

   Process: A series of actions completed to reach a certain goal.

   Output: The amount of something produced.

   Feedback: The response of others to what has been done.

2. Refer to Fig. A. This shows the various parts of a system. It also shows the way in which the parts relate within a system.

3. Refer back to the sketch that accompanies Activity 3 in Section I (page 71). This sketch shows the way in which certain items were identified by the people as resources. These resources, along with certain construction processes, were then used to produce a certain output.

4. Take the two 8½″ × 11″ sheets of paper. Fold each in half lengthwise. Then fold the paper in half widthwise. Fold the paper again widthwise. You will now have folded each sheet into eight panels. Cut or tear these panels apart.

5. On one side of each of the slips of paper, write one of the following terms. Each of these terms identifies one of the items shown in the drawings for Activity 3 of Section I.

   - Bridge
   - Column of rocks
   - The imagined bridge supported by a column of rocks
   - The imagined bridge (without the rock column support)
   - Horse and wagon
   - Hammers
   - Happy people
   - Rocks
   - Saws
   - Stream
   - Unhappy people
   - Wood

6. With your partner, study the drawings for Activity 3 in Section I.

7. Look carefully at each of the system terms just defined. Discuss the part that each of these plays in the story of the bridge-building.

8. Look at the slips of paper one by one. Discuss with your partner whether each of the

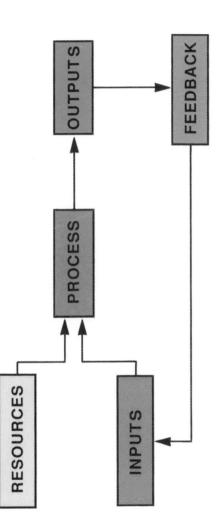

RESOURCES	INPUTS	FEEDBACK	OUTPUTS	PROCESSES
EXAMPLES: TOOLS MATERIALS INFORMATION PEOPLE	EXAMPLES: CUSTOMER NEEDS IDEAS	EXAMPLES: CUSTOMER RESPONSE PRODUCT QUALITY SALES	THE PRODUCT	EXAMPLES: WORK TECHNIQUES

Fig. A. The construction industry may be viewed as a system with various parts. These parts are shown here. Also shown are examples of each of the system parts.

items listed above is an input, a resource, a process, an output, or feedback. In identifying each item by one of these terms, ask yourself what part the item plays in the bridge-building story.

9. When you and your partner have decided on the term (input, resources, etc.) that best describes each of the items listed above, write that term on the back of the slip of paper. Each term may be used more than once. Some of the items shown in the story may not be used at all.

10. When all of the teams have finished, your teacher will discuss the correct answers. Compare your answers with the correct answers and the answers of other teams in the class.

Activity 4: Building a Wind-Powered Generator

Objective

After completing this activity, you will understand the construction and use of a windmill to generate alternating current (AC) electricity.

Materials Needed
- 1 6-volt bicycle light generator
- 4-4″ to 6″ plastic cups
- 1 wheel bearing with ¾″ opening
- 1-10″ × 10″ × ¾″ piece of plywood
- 4-24″ lengths of ¾″ metal conduit
- lengths of ¾″ metal conduit sufficient to construct a generator stand 5′ or taller.
- 15′ of No. 18 electrical wire
- 6-volt light bulb and receptacle
- enough ¾″ aluminum carriage bolts to assemble the generator stand.

Steps of Procedure
1. Figure A shows the general construction of the 6-volt, alternating current (AC) wind-powered generator. The generator used in this activity is available at most stores selling bicycles and bicycle parts. Your assignment

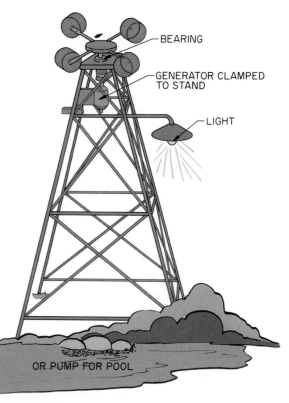

Fig. A. The general construction of a wind-driven generator.

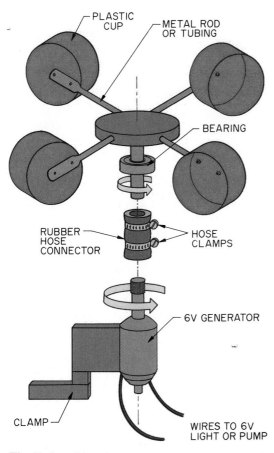

Fig. B. Attaching the generator to the wind-driven section.

is to build a small windmill to turn the bicycle generator. This generator can then be wired to a 6-volt light attached to the stand. It also could power a small 6-volt water pump for a fish pond in your back yard. Its uses are limited only by your imagination.

2. Figure B shows one method for attaching the generator to the wind-driven section of the generator system. The turning shaft of the windmill is attached to the generator with a rubber coupling and two hose clamps.

The bearing is mounted into the wood support of the windmill stand to allow free and easy turning of the windmill shaft. The generator must be clamped to the stand of the windmill so that only the generator shaft turns and not the entire generator.

3. Since the generator must operate outside in all kinds of weather, it is desirable to protect the generator. This can be done by constructing a small plastic shield around the generator. Do not seal it tightly or moisture can still accumulate around the generator and destroy it.

4. Select a location for your generator where there are frequent, steady winds. Since wind gusts can be quite strong, it is best to anchor your windmill stand into the ground or tightly onto the roof to prevent it from being blown over. Embedding the legs into coffee cans filled with cement is one good method of anchoring the stand into the ground.

5. Construction of the stand can be accomplished by flattening and drilling the ends of conduit to bolt the system together. The method you follow in your design may require the use of bolts, screws, and clamps not described in this activity. The main purpose of the stand is to raise the system high enough to expose it to a constant flow of air. At the same time, the stand must be strong enough to withstand heavy wind gusts.

Technical Considerations
- Consider a design large enough to drive a car generator to produce 12 volts of alternating current.
- Consider using a free-turning roof ventilator to generate electricity.

SECTION

IV

THE CONSTRUCTION PROJECT

CHAPTER

9 TYPES OF CONSTRUCTION PROJECTS

Terms to Know

arch bridge
cantilever bridge
cofferdam
commercial
  buildings
dam

frame
  structures
highway
  construction
industrial buildings
mass structures

residential buildings
slab bridge
spillway
suspension bridges
truss bridges

Objectives

When you have completed reading this chapter, you should be able to do the following:

- Describe the three basic types of buildings.
- Identify examples of light, heavy, industrial, and civil construction.
- Identify the basic types of structures.
- Explain the general procedure for constructing a highway.
- Describe the major construction tasks involved in building an airport.
- Explain two general procedures that can be used to construct a tunnel.
- Describe some special construction techniques needed to build a dam.
- Identify and describe five ways in which bridges can be constructed.

Many different types of construction projects are designed and built to meet specific needs in our communities. Houses and apartments, for example, are built to meet our need for shelter. Schools, offices, and stores, in addition to offering shelter, meet many of our cultural and economic needs. Highways, airports, bridges, and dams are built to meet still another need—transportation. In this chapter, you will learn more about various types of construction projects. Fig. 9-1.

TYPES OF STRUCTURES

Buildings can be classified by type. There are two basic types of structures: mass structures and frame structures. **Mass structures** use solid material, such as concrete, for the building's walls. These walls support the building. The weight of the building is carried to its foundation by the walls. **Frame structures** use a frame of metal, wood, or concrete to hold up the building. This frame carries the weight of the building to the foundation.

BUILDINGS

Buildings can be classified into three major types: residential, commercial, and industrial. **Residential buildings** are those in which people *reside*, or live. There are two basic kinds of residential buildings. Those designed to house one family are called *single-family units*. Those designed to house more than one family are called *multiple-family units*. Fig. 9-2. Residential buildings are usually constructed of wood framing or masonry. The foundation varies according to the location of the building. Most foundations are made of either poured concrete or concrete block. Most residential buildings are examples of *light construction*.

Fig. 9-1. Many different types of structures are constructed to meet the various needs of our society. The public transit system is an example of civil construction. The office buildings in the background are examples of heavy construction.

Fig. 9-2. Multiple-family units such as this apartment complex can provide living quarters for many people in a small amount of space.

Commercial buildings are those designed to accommodate businesses. Commercial buildings include buildings such as stores, office complexes, and many types of community service buildings. Fig. 9-3. Large businesses usually have steel-frame buildings on foundations of steel-reinforced concrete. Smaller businesses may have steel-frame, wood-frame, or masonry structures. In some cases, large wood-frame or masonry houses are converted to commercial buildings.

Some buildings are classified as commercial buildings even though they are not used for a specific business. Auditoriums, churches, convention centers, schools, libraries, and courthouses are commercial buildings. Such buildings are built in the same way as other commercial buildings from the same kinds of materials. Fig. 9-4. Some commercial buildings are examples of *heavy construction*.

Industrial buildings house the complex machinery that is used to manufacture goods. Industrial buildings are generally low buildings of only one or two stories. Nevertheless, this type of building can be huge, spreading out over a large amount of land. Some industrial buildings

Fig. 9-3. Some commercial buildings, such as the famous World Trade Center, provide offices for thousands of businesspeople.

Fig. 9-4. A school is considered a commercial building.

cover several acres of land. Industrial buildings are examples of *industrial construction*.

Processing industrial materials is a large-scale operation that requires special kinds of manufacturing facilities. In some respects, construction of these facilities is similar to other types of construction. The basic steel framework for industrial buildings must be erected. Both large and small buildings must be built, and utilities such as plumbing and wiring must be installed. However, in industrial plants, large machines and special equipment must be installed. Special foundations are required for the heavy machines and equipment.

Each different type of industrial plant has different construction needs. Fig. 9-5. For example, steel mills and petroleum refineries need

Fig. 9-5. A petroleum refinery is one example of an industrial building that has special construction requirements.

special furnaces. Petroleum refineries also need complex fractionation towers to separate crude oil into its different components, such as gasoline and diesel fuel. Chemical processing plants also have special requirements. They need chemical baths and storage buildings that are highly resistant to chemical erosion.

For Discussion

This section describes the three types of buildings, residential, commercial, and industrial. In your town or city, do you see any examples of residential buildings that have been converted to a commercial use? What types of businesses do such buildings usually house?

||| HIGHWAYS

Highways, airports, tunnels, bridges, and dams are discussed on the following pages. These are examples of *civil construction*.

Highways are constructed to give wheeled vehicles a proper surface on which to travel. **Highway construction** is a general term used for the construction of any road or street. The basic steps of highway construction are preparing the soil, preparing the roadbed, and striping the finished road.

To prepare the soil for highway construction, trees, roots, and rocks must be removed. Then the ground must be leveled and graded to the proper elevation. Earth must be removed from areas that are too high, and low spots must be filled in. Fig. 9-6.

Next the roadbed is prepared. There are two major types of roadbeds: flexible and rigid. A flexible roadbed requires a thick gravel subbase. The subbase spreads the load on the highway into the soil beneath the highway. It is usually covered with a concrete base. The top layer, a mixture of sand, asphalt, and either gravel or

Fig. 9-6. Building a road requires extensive earthworking.

crushed rock, is designed to give slightly under a heavy load. This helps protect the surface of the highway from cracks and potholes due to stress.

A rigid roadbed consists of steel bars set on a sand base. Approximately 8 inches (20 cm) of concrete are poured over the bars. The resulting steel-reinforced concrete slab is capable of spreading the weight of traffic over a large area.

Before a new road can be used, it must be striped. Concrete pavement can be striped as soon as it hardens. Bituminous pavement has to cure for a week or two before it is ready to be striped.

For Discussion

This section describes two types of roadbeds: flexible and rigid. Based on what you have read in the book, what type of roadbed lies beneath the road that runs in front of your school?

AIRPORTS

The construction of an airport is actually a combination of road construction and building construction. The major road construction tasks for an airport include construction of the following:

- runways
- taxiways (the paths between the passenger terminals and the runways)
- aprons (the areas near the passenger terminals where planes park for boarding and deboarding)
- parking lots and roadways to handle the cars people drive to the airport

These four tasks may require many miles of road surfaces. For example, large international airports must have runways long enough to allow a Boeing 747 to land. Such a runway must be 13,000 to 14,000 feet (3,900 m to 4,200 m) long. The runways, taxiways, and aprons must also be strong enough to support the 350-ton aircraft. This requires a strong soil base and thick, steel-reinforced concrete. Fig. 9-7.

Fig. 9-7. The paved surfaces at airports must be made to be extra heavy-duty. The concrete for some runways is more than 10 feet (3 m) thick.

HEALTH & SAFETY

There are types of pollution other than air pollution and water pollution. There is also noise pollution. Noise pollution can have a serious effect on health. It can, for example, contribute to stress. As air traffic has increased, the number of airports has grown. People living close to airports have been subjected to the sound of arriving and departing aircraft. A number of things have been done to reduce airport noise pollution. Many of these concern aircraft takeoff and landing procedures. New engine designs also have been introduced. These new designs have been able to cut some of the engine noise. Engineers are working on other ways to reduce aircraft engine noise.

Airports also need several kinds of buildings. Passenger terminals, freight terminals, hangars, control towers, maintenance buildings, and fire stations are all important to the operation of an airport. Each of these structures has special construction requirements. For example, a hangar must be large and strong enough to house one or more planes. It must be designed to give mechanics easy access to the planes. Fig. 9-8. A control tower must be designed to give air traffic controllers a clear view of the runways and all the air traffic in the area. Most control towers have huge windows on all sides to meet these needs.

Specially trained airport planners and engineers make the plans for most large airports. They work with city officials to find the best site for the airport. They study population trends to make sure the airport is built in a location that will be useful for many years. They determine the size of the airport according to the amount of air traffic it is expected to handle. Together with the city officials, they design an efficient system of buildings and runways. Because large airports require a long time to build, they are usually built in stages. As each stage is finished, it is opened for use by the public.

Fig. 9-8. Hangars are designed and constructed to give mechanics access to the planes for maintenance and repairs. This plane is being prepared for repainting.

Did You Know?

One of the major construction features of an airport is the runway. The earliest airports had only a single runway. Later, as air traffic increased, the number of runways also increased. Some airports had four and five runways. These various runways were needed because planes were relatively light. They could be affected by crosswinds on landing and takeoff. Thus, they needed to take off and land along runways that would minimize the effects of crosswinds. In the last twenty years, aircraft engines have become more powerful. Aircraft also have become heavier. This has meant that aircraft are not as affected by crosswinds. Many major international airports have only two main runways.

For Discussion

The text mentions several factors that must be considered in finding the best site for an airport. Can you list others not included in the text?

||||TUNNELS

Tunnels are constructed as underground passageways for roads and railroads. Their usual purpose is to streamline traffic or to route traffic around or through an obstacle. Figure 9-9 shows a modern rapid transit system that uses tunnels to route the tracks under parts of the city.

Fig. 9-9. Much of the Metro Rapid Transit System in Washington, D.C., is routed through tunnels under the city.

Sometimes tunnels are built under rivers. Such tunnels enable highway traffic to go under the rivers so that the rivers can stay open for ship traffic. Tunnels for roads and railroads are made through mountains because the route through the mountain is the shortest route to the other side.

Special tunnel-boring machines are used to create tunnels in soft rock such as limestone. Fig. 9-10. They use disk-shaped cutters to dig mud, cut through rock, and move loose earth out of the way. A steel *shield* is used to keep the earth around the tunnel from caving in until workers can stabilize it. Workers stabilize the walls of the tunnel by lining them with concrete or cast iron.

Sometimes tunnels must be built through hard, solid rock. In these cases, explosives are used to *blast*, or break up, the rock. Engineers must study the rock carefully to determine the proper amount of explosives. A small portion of rock is blasted at a time. Then the loose rock is removed and steel beams are inserted to support that portion while the next portion is blasted. Some tunnels built in hard rock are strong enough that they do not need iron or concrete supports. However, most are eventually lined to make them more durable.

Did You Know?

In 41 A.D., the Romans constructed a tunnel to drain a lake in central Italy. It was an enormous project, taking ten years and requiring the work of over 30,000 men. Today, tunnels are constructed using large tunneling systems. These tunneling systems consist of a large boring head, a power unit, and machinery for erecting the parts of the tunnel lining. There is also a conveyor for removing the material that has been cut away. These tunneling systems have allowed the construction of longer and larger tunnels. Such tunneling systems have also allowed tunnels to be completed more quickly.

Fig. 9-10. Special tunnel-boring machines use disk cutters to cut tunnels through soft rock.

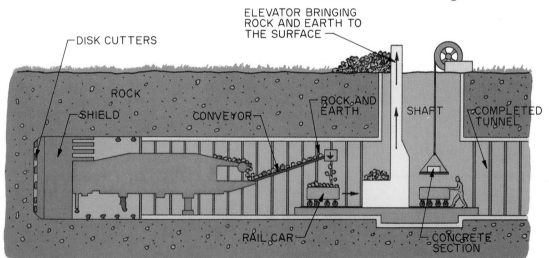

For Discussion

Tunnels are more common in regions where there are mountains and rivers. In or near your city or town, are there any tunnels that are used for traffic?

Fig. 9-11. This pedestrian bridge allows school children to cross a busy highway safely.

||| BRIDGES

A bridge is a structure that is built to span, or cross over, a river or a gap in the earth. A bridge provides a way for people and vehicles to cross from one side to the other. Bridges can also span other structures. A bridge on a highway might cross over railroad tracks or over another highway. Different types of bridges carry railroads, highway traffic, pipelines, and foot traffic. Fig. 9-11. Sometimes bridges are made with movable sections that can be raised or swung out of the way so that large boats can pass by. Fig. 9-12.

Bridges can be constructed in several different ways, depending on the required length of the bridge and on the weight it must support.

Fig. 9-12. This bridge can be raised to allow barges to pass through.

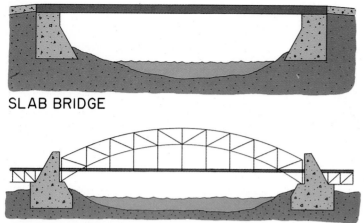

SLAB BRIDGE

ARCH BRIDGE

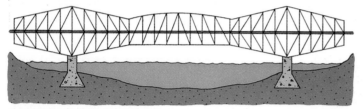

CANTILEVER BRIDGE

Fig. 9-13. Different kinds of bridges: slab, arch, truss (two types), cantilever, and suspension.

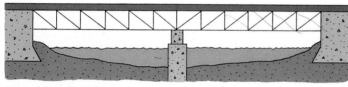

TRUSS BRIDGE

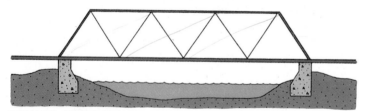

TRUSS BRIDGE

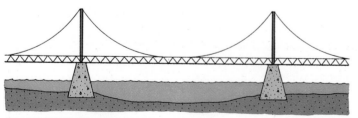

SUSPENSION BRIDGE

Fig. 9-13. Most bridges are anchored on each end by supports called abutments. The simplest type of bridge, the **slab bridge**, consists of a concrete slab supported by abutments. Some of the longer slab bridges are also supported by a pier, or beam, in the middle. This type of bridge is usually made of steel or steel-reinforced concrete and is used mostly for light loads and short spans.

Did You Know?

Computers are used to solve many design problems. This is especially true in the construction industry. For example, in designing suspension bridges, engineers have been able to calculate the shape of the suspension cables under changing loads.

An **arch bridge** is one in which an arch is used to carry the weight of the bridge. Arch bridges are made of concrete or steel and are usually constructed over deep ravines. **Truss bridges** are supported by steel or wooden trusses, or beams that are put together to form triangular shapes. Triangles are used because the triangle is a particularly strong structural shape. Trusses are also used in combination with other bridges to give them additional support. One type of bridge that commonly uses trusses is a cantilever bridge. A **cantilever bridge** is used for fairly long spans. It has two beams, or cantilevers, that extend from the ends of the bridge. They are joined in the middle by a connecting section called a *suspended span*. The whole structure usually receives additional support from steel trusses.

The very longest crossings are spanned by suspension bridges. **Suspension bridges** are suspended from cables made of thousands of steel wires wound together. One well-known example of a suspension bridge is the Golden Gate Bridge in San Francisco. Fig. 9-14.

Fig. 9-14. The Golden Gate Bridge in San Francisco is an example of a suspension bridge.

Did You Know?

The ancient Romans were among the greatest bridge builders in history. They developed a type of natural cement that could be used to bond stones together. They also developed the cofferdam. This is a watertight enclosure in which the piers, which support the bridge, can be built. The Romans also took full advantage of the circular masonry arch. Examples of the skill of the Romans in building bridges can still be seen. Some of the bridges they built are still in use.

For Discussion

The text discusses the main types of bridges. Are there any bridges in your town or city? What types of bridges are they?

|||| DAMS

A **dam** is a structure that is placed across a river to block the flow of water. This is usually done for one of two reasons. The most common reason for damming a river is to create a water reservoir for nearby communities. Fig. 9-15. Another important reason to dam a river is to collect water to power the water turbines in a hydroelectric power station.

Because a dam has a major effect on the surrounding environment, the site for the dam must be selected very carefully. Civil engineers typically spend years studying the characteristics of the land near a proposed dam site. The reservoir created by the dam may flood land that was previously inhabited or used for farming. Arrangements must be made to relocate people and structures that will be flooded. Fig. 9-16.

The construction of a dam is a major undertaking. A large dam may take ten years or longer to construct. A tremendous amount of earthworking must be done before construction on the dam can even be started. The soil must be made strong enough to hold both the dam and the weight of the water in the reservoir created by the dam. The earth must be built into an embankment to keep water from spilling around the dam. The dam itself must have a solid, strong foundation so that it will not wash away when the river swells. The dam must have strong gates to allow water to pass through the dam in controlled amounts.

A spillway must also be constructed. A **spillway** is a safety valve that allows excess water to bypass the dam. This is necessary because few dams are strong enough to withstand the force of floodwater. If the water could not bypass the dam, the dam would break.

Many underwater parts of the dam must be built on dry ground. To be able to work on dry ground, the construction workers must divert the river. This alone can be a major project if the river is a large one. A **cofferdam**, or watertight wall, must be built to keep water out of the worker's way. Fig. 9-17. This temporary wall can be made of timber, concrete, soil, or sheets of steel.

As work on the dam progresses, another cofferdam is built farther out in the river, and the first cofferdam is removed. This allows the water to flow around the sides of the construction site so that the river is not obstructed. After the workers finish the last underwater portion of the dam, the cofferdam is removed entirely.

Fig. 9-15. The Hoover Dam, also known as Boulder Dam, is a dam on the Colorado River. The lake formed behind the dam is Lake Mead.

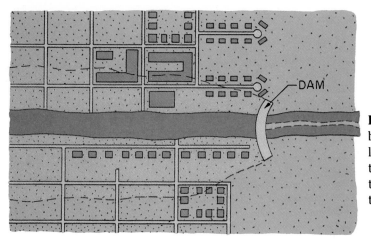

Fig. 9-16. The area shaded light blue in this illustration represents land that will be submerged after the dam is built. People who live in this area must make arrangements to move.

RIVER BEFORE DAM

RIVER AFTER DAM

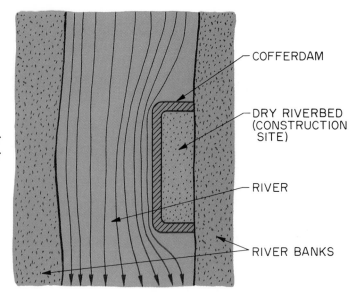

Fig. 9-17. A cofferdam diverts the river so that workers can build the underwater parts of a dam.

COFFERDAM

DRY RIVERBED
(CONSTRUCTION
SITE)

RIVER

RIVER BANKS

Did You Know?

The designers of some dams have considered the needs of migratory fish, such as salmon. For example, adult salmon will try to swim upstream to their spawning grounds. Young salmon will try to swim downstream, away from the spawning grounds. A dam will prevent this. To allow the fish to swim past the dam, fish passes have been built at some dams. These sometimes take the form of fish ladders. A fish ladder consists of a series of small pools. The fish move from one pool to the next. Such conservation measures have been introduced in Canada, the northwest United States, and Scotland.

For Discussion

Are there any dams in or near your community? Are the dams used to create a water reservoir or to power hydroelectric turbines? Are the dams used for both?

Construction Facts

TENN-TOM HEAVY CONSTRUCTION

In late 1986, Alabama and Mississippi celebrated the opening of a 234-mile (375-km) waterway. This mammoth waterway connects northeastern Mississippi to the gulf coast of Alabama. It connected the Tennessee and Tombigbee rivers to open up a whole new commercial waterway in the southeastern United States. Its name, appropriately enough, is the Tennessee-Tombigbee Waterway, or Tenn-Tom.

Tenn-Tom is the largest marine construction project ever attempted in the United States. This massive earthmoving job was begun in 1972. Three hundred million cubic yards of earth were moved to prepare the waterway.

To meet the construction schedule, Tenn-Tom constructors spent $43 million on specially designed equipment and put their workers on a three-shift-per-day schedule. One interesting piece of heavy equipment used on the job was a specially-built excavator-loader. Powered by a pair of large bulldozers, the machine could fill a 50-ton dump truck in less than one minute.

One problem with this massive excavation was what to do with the excavated earth. Thirty-eight different disposal sites, located in deep valleys, were the solution. The excavated soil was spread over 5,000 acres (2000 hectares) of land.

CHAPTER 9

REVIEW

Chapter Summary

Many different types of construction projects are designed and built. These include residential buildings, commercial buildings, and industrial buildings. Highway construction is a general term for the construction of a road or street. The construction of an airport is a combination of road construction and highway construction. Tunnels are constructed as underground passageways. They require special construction techniques. Bridges can be constructed in several different ways. There are several types of bridges: slab, arch, truss, cantilever, and suspension. A dam is a structure placed across a river to block the flow of water. A cofferdam enables the builders of the dam to work on dry ground.

Test Your Knowledge

1. What are the three basic types of buildings?
2. Give two examples of special construction needs that an industrial plant might have.
3. From what two types of concrete can pavement for highways be made?
4. Name four examples of road construction that must be done to build an airport.
5. Who is responsible for planning and designing most large airports?
6. What technique is used to remove hard rock from a tunnel site?
7. What type of bridge is most commonly used for short spans and light loads?
8. What type of bridge is used to span the longest crossings?
9. What is the purpose of a spillway?
10. What is a cofferdam?

REVIEW

Activities

1. Keep a log of construction projects you see being built. Make notes about what you see happening that interests you. Report your findings to your class.

2. On a sheet of paper, make four columns. Label each column with one of the four major types of construction projects. Plan a trip around your town or city to find and look at the various structures that have been built. In the appropriate column, write a short description of the type of structure. See how many of each type you can find.

3. Select an interesting structure such as one of the following:
   - Gateway Arch
   - Washington Monument
   - Crystal Palace
   - Golden Gate Bridge
   - Great Wall of China
   - Egyptian Pyramids
   - Machu Picchu

   Do library research on your selected structure. Write a one-page report on how it was built. Write also about the human needs that prompted the building of the structure. Report your findings to your class.

CHAPTER **10**

THE DECISION TO BUILD

Terms to Know

alterations	power of eminent
bond	domain
feasibility study	private project
interest	public project
letter of commitment	renovation
mortgage note	zoning laws

Objectives

When you have finished reading this chapter, you should be able to do the following:

- Identify two construction-related alternatives to building a new structure.
- Describe the difference between private and public building projects.
- Describe the process of selecting and acquiring a building site.
- List three sources of funds for construction projects.

Every construction project begins with a decision. Someone decides that an existing structure is inadequate. Once that decision is made, many other decisions are necessary. Is a new structure needed, or can an old structure be modified to meet the need? The answer to this question determines what kind of construction work will be necessary. For example, sometimes the structure can be modified to meet the owner's needs. Other structures may present problems that are not easily solved. In these cases, the best solution may be to build a new structure.

MODIFYING AN EXISTING STRUCTURE

Modifying an existing structure can sometimes save the owner a lot of money. Even adding a room onto a house is usually much less expensive than building a new house. Many people therefore decide to modify an existing structure rather than build a new one. Fig. 10-1.

Before modification can begin, the owner must decide what should be modified and the extent of the modification. Some modifications are minor, such as knocking out a wall to combine two rooms, or installing a skylight. Others are major, such as adding a room onto a house. Modifications also include additions such as patios. Fig. 10-2.

The type of modification needed depends on the present condition of the structure. Some owners need to make design alterations to meet their changing space requirements or to increase the attractiveness of the structure. Other owners may want to update part or all of a structure to provide more modern comforts and conveniences.

Alterations

Alterations are changes in the structural form of a building or other structure. These changes are usually made to make the structure more attractive or useful. If the owner of a building decides that the building is no longer adequate for its purpose, he or she may choose to *remodel*, or change, the existing building. A family may decide to add another bedroom to make room for a growing number of family members. The owners

Fig. 10-1. Buildings are sometimes renovated to meet other needs. This tropical fish store was once a gas station.

Fig. 10-2. Here a patio and family room addition are being added to an existing home.

of office buildings, supermarkets, and stores frequently remodel their buildings to make them more attractive. Factories are often remodeled to make room for new equipment and machinery.

A major remodeling project may be quite expensive, but it usually costs less than building a new structure. Some contractors specialize in remodeling. These companies do not build new structures. They are experts at remodeling buildings. Most remodeling contractors can examine a proposed remodeling project and give you an idea of how much it will cost to complete the project.

Renovation

Sometimes people who own older buildings find that the buildings no longer meet their needs. However, many of these buildings have attractive characteristics that cannot easily be duplicated. For example, an old house may have a beautiful oak staircase or 10-foot ceilings. The owner of such a building may choose to renovate it.

Renovation is the process of restoring the original charm or style of the building, while at the same time adding modern conveniences such as air conditioning. Fig. 10-3. Often the plumbing and wiring systems also need to be replaced. Because of the nature of the problems that must be fixed, renovation may be a major project. Sometimes renovation even costs as much as new construction. However, the owner is able to keep the attractive feature, such as an oak staircase, without having to pay a high price to have them specially built.

Many cities are now attempting to renovate whole districts that contain old buildings. Some of these cities offer the buildings to potential buyers at an extremely low cost. One condition of these sales is that the buyers renovate the buildings within a certain time period. In this way, a city can clean up its downtown areas. At the same time, it can save historical buildings in these areas—all at a very low cost to the city.

Fig. 10-3. This house is being renovated. Renovation allows its owner to have a comfortable home and, at the same time, preserve a historical building.

HEALTH & SAFETY

Older homes are sometimes renovated to correct possible hazards. There have been a number of improvements in home construction in the last twenty years. As one example, home wiring systems have been improved. New insulating materials have been developed. Often, renovation in an older home may involve the replacement of outdated and perhaps dangerous electrical wiring. It might also involve the replacement of dangerous asbestos insulations with other insulating materials.

For Discussion

You may be familiar with the processes of alteration and renovation. Perhaps there are houses in your neighborhood that have been altered or renovated. Perhaps you have seen the work being done. Why do you think the changes were made to the buildings?

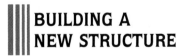

BUILDING A NEW STRUCTURE

If the owner decides that altering or renovating an existing structure will not solve the problem, he or she probably needs to build a new structure. After the need for the new structure is established, many more questions need to be answered. For example, what type of structure is needed? Who will pay for the construction? When, where, and how will the structure be built? These and other questions are answered by the owner or owners of the project.

Some construction projects are *privately owned*. This means that they are owned by an individual person or company. For example, most houses are privately owned. Other projects are *publicly owned*. They are paid for with taxes and belong to the public. School buildings and public libraries are common examples of publicly owned projects. Fig. 10-4.

Private Projects

When a construction project belongs to an individual or to a company, it is known as a **private project**. Some private projects are planned to meet the needs of a company or business.

Individual Ownership

Individually owned projects are usually smaller than public projects. Fig. 10-5. The addition of a deck to an existing house is an example of a small, individually owned project. Even the construction of a new house is a relatively small project.

When you first get the idea to build, you will need to ask yourself many questions. As an example, pretend that you want to build a new home for your family. You will need to begin by asking yourself the following questions:
- Do we really need (or want) to build this house?
- Could we add another room onto our old house instead of building a new house?
- How much better will a new house meet our family needs?

If you do decide that you need or want to build a new house, then you must ask more questions and make more decisions.

Fig. 10-4. A school is one type of public project.

Fig. 10-5. Adding a deck to a house is an example of an individually-owned project.

- Can we afford to build a new house?
- What financing is available to help us pay for building a house?
- Can we sell our old house for a fair price?

The answers to these questions are critical. No matter how much you may want, or even need, a new house, you must approach such a project practically. If you can give satisfactory answers to these first six questions, then you can go on to define the details.

- What type of design do we want our new house to have?
- Should we build in town, near schools and shopping center, or in the country?
- What land is available, and at what price?

Corporate Ownership

Many commercial buildings are owned by corporations. Since these buildings are owned by the corporation, they are constructed as private projects. Common examples of private projects are office buildings, hospitals, warehouses, stores, and factories. Fig. 10-6.

In the case of a corporation, the decision to construct a new building is made by the board of directors. The board of directors must study the proposed construction from every point of view. Is the building necessary? If so, what type of building would best suit the needs of the corporation? Where is the best place to build the building? Does the corporation need to buy more land or can the building be built on property the company already owns? How will the corporation pay for the building? The company could waste a lot of money by making a wrong decision about any one of these questions. The answers must therefore be carefully thought out before a company can decide to build.

Fig. 10-6. This factory building was built to meet a company's need for manufacturing space.

Public Projects

Some projects are paid for with tax money. These projects belong to the whole community and are known as **public projects**. Public projects are planned to meet the needs of the community. Fig. 10-7. Some of the projects are rather small. A new sidewalk in a city park and minor road repairs are examples of relatively small public projects. These projects may cost only a few thousand dollars. A new interstate highway is considered a large project. A new airport is also a large project. These projects may cost millions of dollars each.

Some public projects take a long time to get started because there are so many people involved in the initial, or first, decision-making process. Sometimes it is necessary to pass around a petition and get signatures from taxpayers who show their support for a project. Fig. 10-8. In some cases, people have the opportunity to vote for or against a project. Sometimes a public hearing is called so the proposed project may be discussed publicly. Fig. 10-9.

Fig. 10-7. This highway and the bridge in the background were built to meet the public's need for better transportation facilities.

Fig. 10-8. Sometimes taxpayers sign a petition to show their support for a proposed public project.

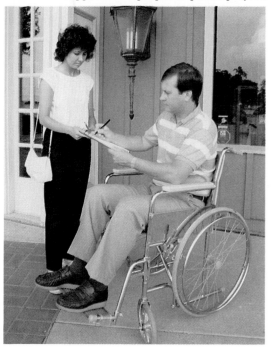

Fig. 10-9. A public hearing allows people to speak for or against a proposed project.

Did You Know?

The sites of many cities have been inhabited for hundreds of years. Often, when the foundations are being dug for a new building, workers unearth items from another time. When this happens, construction may be halted so that the site can be studied by archeologists. These are scientists who study the way humans lived a long time ago. They do this by studying their tools and weapons. Construction may be delayed to allow archeologists to study a building site more closely. If the site is very important in the eyes of the archeologist, building may be halted, or even called off. In other instances, there may be an attempt to excavate a part of the site and perhaps open it to public display.

For Discussion

The book discusses projects that are privately owned and publicly owned, can you identify some publicly owned buildings in your town or city? What are they used for?

BUILDING SITES

As you learned in Chapter 6, the location of a construction project is called a _site_. The site is an important part of any construction project. The type of foundation used for a building, for example, depends on the type of soil and, to some extent, the slope of the land. The owner must either choose a site that is appropriate for the proposed structure or design a structure that is appropriate for the site. Because choosing a site is a critical part of a construction project, gathering information about the site is part of the feasibility study. Many potential sites are studied to find the one that is best suited to the project.

Selecting the Site

Much of the success of the building project depends on making a wise site choice. Individual owners are responsible for choosing their own building sites. In private industry, managers and the board of directors must approve the site. For public projects, government officials make the site decision. Fig. 10-10.

Zoning Laws

When selecting a suitable site, the owner, directors, or officials must consider the local zon-

Before construction can begin on a public project, a great deal of planning must be done. Many questions must be answered. The first questions are the same as those for a private project. Is there a need for the project? How will the project benefit the community? How will the project be financed?

After the initial decision has been made to build a project, a **feasibility study** is done to gather information about the proposed project. Good information is important because it helps people make good decisions. A feasibility study helps the decision-makers decide if it is feasible, or practical, to build the project. It includes information on the cost of the project and financing options as well as the availability of land and essential materials.

Fig. 10-10. Aerial photos such as this one can help owners or government officials select the best site for a new project.

ing laws. Cities are divided into residential, business, and industrial zones. The site for the project must be in the proper zone. **Zoning laws** tell what kinds of structures can be built in each zone. Fig. 10-11. Residential zones are areas reserved for homes. Factories and warehouses are located in industrial zones. Industrial areas are set apart because the noise, smoke, and traffic of industries can be unpleasant in a residential area. Fig. 10-12. Commercial zones are usually located near residential zones. This allows people to get to stores and offices easily.

Did You Know?

Zoning laws were introduced in some German and Swedish cities about 1880. These laws applied to land that was being built on just outside the city. Most of the zoning laws at this time were concerned with regulating the height of buildings. In the United States, most zoning laws are concerned with regulating the use of land for a certain purpose. Thus, some locations are zoned "commercial," or suitable only for businesses. Other locations are zoned "residential," or suitable only for houses and apartments.

Other Considerations

Some of the other factors that must be considered are the location, cost, and characteristics of the site. Fig. 10-13. Land that is in the city, close to schools and shopping centers, is generally more expensive than property in the

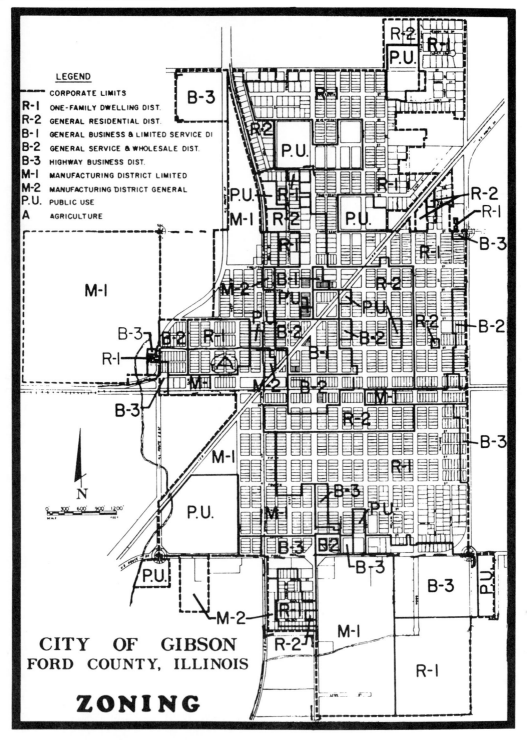

Fig. 10-11. The zoning map indicates the various zones in a city.

Fig. 10-12. This area is zoned *industrial* to protect residential property owners from noise, smoke, and traffic.

Fig. 10-13. In selecting a site for construction, many factors must be considered. Among these factors are location, cost, physical characteristics, site considerations, utilities, zoning, and accessibility.

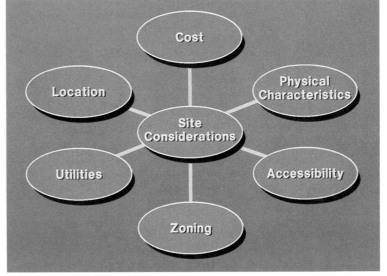

country. Accessibility is an important factor, especially for residential and commercial buildings. Is the land easy to get to? Are the roads paved? Fig. 10-14. Utilities can be a critical issue if the property is too far out in the country for city water and sewers.

All the physical characteristics of the site, such as hills, trees, and soil conditions, must be considered. Is the land suited to the type of structure to be built? How much earthworking must be done to achieve the level, stable conditions necessary for building a structure? All of these things need to be evaluated carefully to be sure that the most desirable site is selected.

Did You Know ?

A wetland is land that contains much soil moisture. Swamps, bogs, and marshes are wetlands. In the past, wetlands often were drained. The land could then be used for farming or for building sites. Wetlands, however, provide a refuge for certain birds. They also allow a variety of interesting plants to grow. Today, there is an effort to preserve wetlands. A building site that is partially wetland is often carefully evaluated. Often, in the interests of conservation, construction may be limited to that part of the site that is not judged to be wetland.

Fig. 10-14. A paved road helps make a site easily accessible.

Acquiring the Site

Once the site is selected, steps must be taken to acquire (get) the property. Sometimes this is an easy procedure that takes only a few days. Generally, however, acquiring the site is an involved process that takes a long time to complete.

Individually Owned Projects

Property for private projects is usually acquired by negotiation. *Negotiation* is the process of working out an agreement between the buyer and the seller. Imagine that you have a bicycle for sale and someone wants to buy it. The two of you will have to negotiate a final sale price with which you will both be happy. Then the sale can take place.

Negotiations for land are usually handled by a real estate agent. A real estate agent is someone who specializes in helping people find and purchase suitable property. The real estate agent helps the buyer and seller come to an agreement. Fig. 10-15.

Once the terms of the sale are agreed upon, the sale must be made legal. When the buyer and seller agree, they sign a contract. A contract is an agreement in writing that states the terms of the sale. The written agreement ensures that both the buyer and the seller know exactly what is being sold and for what price. Then the money changes hands and the land sale is recorded in the county records. At this point, the ownership of the site is legally changed.

Corporate-owned and Public Projects

The procedures for buying property for public projects and for large commercial projects are similar to those for buying individually-owned property. One difference is that the terms of the contract are usually much more complex.

Fig. 10-15. A real estate agent handles the negotiations between seller and buyer.

Another major difference is that an attorney is usually contacted instead of or in addition to a real estate agent. By using an attorney, the buyers make sure that all of the complicated terms of the contract are legal.

Sometimes several *tracts*, or pieces, of land are needed for a public project. Fig. 10-16. Some of the land may be owned by someone who does not want to sell. However, through the **power of eminent domain**, the government has the right to buy the property for public purposes even though the owner does not want to sell. The government *condemns*, or takes, the property in the interest of the general public. Fig. 10-16. The owner is given a fair price for the land, so he or she is not cheated, only inconvenienced.

Fig. 10-16. This notice states that the property has been condemned. The building will be demolished.

For Discussion

The book discusses the need for carefully selecting a building site. All building sites are not level. In your town or city, can you point out ways in which a building has been designed for a building site that is not level?

FINANCING

After the plans for a building project have been completed, the method of financing the project must be determined. If the project is privately owned, the money is usually borrowed. For a public project, the money comes from taxes or bonds.

Mortgages

Very few people have enough cash to pay for an entire building project. The money must be borrowed from a lending institution such as a bank or a savings and loan association. First, a loan application must be submitted. Fig. 10-17. The bank checks the income and the credit rating of the person or group applying for the loan. The bank needs to know whether the applicant has the ability to pay back the loan.

Letter of Commitment

Once the loan is approved, the lending institution prepares a **letter of commitment**. This states the terms and conditions that they require for payment of the loan. This letter of commitment states the loan's interest rate and the number of payments in which the borrower must pay back the loan.

The letter of commitment is important both to the lender and to the applicant. It states the specifics of the loan so that the applicant is not surprised at the last moment. The applicant knows in advance whether he or she can make the payments comfortably. For the lender, the

Fig. 10-17. The information on the loan application helps the lending institution decide whether to loan money to the applicant.

letter of commitment is a guarantee that the applicant is serious about wanting a loan and is willing to risk a penalty if he or she backs out at the last minute.

Mortgage Note

The next document that is signed is called a **mortgage note**. The mortgage note is actually two documents. The *note* is a document in which the lender agrees to finance a project under certain conditions and at a specified interest rate. The *mortgage* pledges the property as security for the loan. In other words, if the borrower does not make the monthly payments, the property becomes the property of the lending institution that holds the mortgage.

Interest

All lending institutions charge interest on the money they lend. **Interest** is the price the borrower pays for using the lending institution's money. The rate of interest is usually expressed as a percentage of the total amount of the loan. For example, suppose that a person borrowed $100 for one year at a yearly interest rate of 10 percent. At the end of the loan period the borrower would owe the lender the $100 plus 10 percent ($10), or a total of $110. Interest makes lending worth the lender's risk and effort to make the loan. The lender decides what the interest rate will be.

Because of interest, a *mortgagor*, or borrower, actually pays much more for the property than the original selling price. However, without a mortgage, many people would never be able to own property and have a home. Most mortgage loans are made for a period between ten and thirty years. Thirty years may seem like a long time to be paying for a house, but the longer the time period is, the lower the monthly payments are. Many people prefer to stretch their house payments over many years. However, the faster the loan is repaid, the less the total cost of the property will be.

Did You Know?

It is important that mortgages be recorded. Unless there is a record of a mortgage, it would be possible for the person who has taken out the mortgage (the mortgagor) to take out another mortgage with another person. Thus, the mortgagor would be borrowing twice using the same property. By law, on the continent of Europe, mortgages have had to be recorded since the eighteenth century. In North America, their recording has been required since the nineteenth century.

Bonds

Bonds are usually used to help finance commercial and public projects. A **bond** is a type of note by which various amounts of money are borrowed from many different people or institutions at once. Fig. 10-18. The purpose of selling bonds is to obtain money for construction. For example, school districts sometimes sell bonds to finance a new school. Bonds are a common way of financing public projects such as schools, highways, and water treatment plants.

Bonds can be sold to individuals or to lending institutions. The seller pays interest on the bond over a certain number of years, called its *lifetime*. The lifetime of a bond may be 5, 10, or 20 years. In addition, the seller promises to repurchase the bonds for their face value. The *face value* is the stated amount of the bond. For example, a $1,000 bond has a face value of $1,000.

Appropriations

Public projects may also be financed with money collected as taxes. Federal, state, and local governments *appropriate*, or set aside, tax money for particular construction projects. The appropriation money is usually placed in a special account at a bank. This money is used to pay for the site and for all the construction costs. Fig. 10-19.

For Discussion

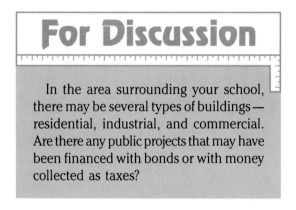

In the area surrounding your school, there may be several types of buildings—residential, industrial, and commercial. Are there any public projects that may have been financed with bonds or with money collected as taxes?

Fig. 10-19. Tax money was used to pay for the construction of this space shuttle launch site.

Fig. 10-18. People buy bonds as an investment. The bonds are redeemable with interest on a specified date.

Construction Facts

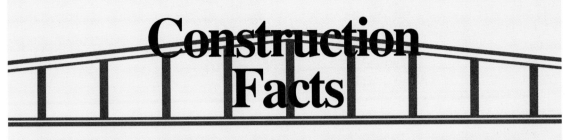

CHUNNEL UNDER THE CHANNEL CONNECTS ENGLAND AND FRANCE

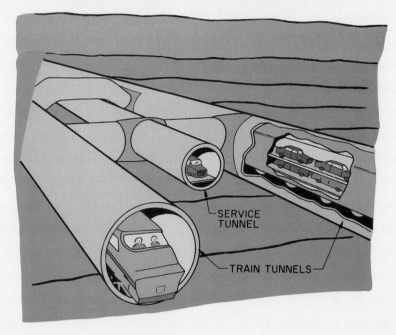

SERVICE TUNNEL

TRAIN TUNNELS

For over 200 years, engineers have dreamed of a tunnel beneath the English Channel that would connect England and France. Now it looks as if the "Chunnel" may one day be a reality. In 1987 work began on what is the largest civil engineering project in the history of Western Europe.

The Chunnel consists of three parallel tunnels. Two of the tunnels will be for double-deck railroad trains carrying passengers, automobiles, trucks, and buses. The trains will travel at speeds up to 100 mph. The smaller middle tunnel will be used for service and ventilation.

The tunnels will be about 300 feet below the surface and thirty-one miles long. Twenty-three of those miles will be under the water.

Both England and France are working on the Chunnel, which requires about 1000 workers.

The Chunnel is scheduled for completion in 1993. A second tunnel involving drive-through auto tunnels has been proposed for the year 2000.

CHAPTER 10

REVIEW

Chapter Summary

Modifying an existing structure can sometimes save money. Modifications will depend on the condition of the structure. Alterations are changes in the structural form of a building. Renovation is the process of restoring the original style of a building, while adding modern conveniences. Some construction projects are privately owned. Others are publicly owned. A feasibility study gathers information about a proposed building project.

The location of a construction project is called a site. Zoning laws tell what kinds of structures can be built on each site. Acquiring the site can sometimes be an involved process. Negotiation is the process of working out an agreement between the buyer and the seller. A contract is a written agreement that states the terms of the sale. The power of eminent domain allows the government to buy property for public purposes even though the owner does not want to sell.

A letter of commitment states the terms and conditions a lending institution requires for payment of a loan. A note is a document in which the lender agrees to finance a project under certain conditions and at a specified lending rate. A mortgage pledges the property as security for a loan. Interest is the price the borrower pays for using the lending institution's money. A bond is a type of note by which various amounts of money are borrowed from many different people and institutions at once.

Test Your Knowledge

1. What are the two main types of modifications that can be made to a structure?
2. What are the two main types of project ownership?
3. What is the purpose of a feasibility study?
4. What is the name of the laws that control the locations at which different kinds of structures can be built?
5. Name three factors that must be considered before a site is purchased.
6. What document is a written agreement stating the terms of a sale?
7. What is the power of eminent domain?
8. How is money usually obtained to finance privately-owned projects?
9. In what two ways can a public construction project be financed?
10. What is a bond?

REVIEW

Activities

1. Sketch a plan of an alteration you would like to have made to your house or apartment. Explain how the alteration would solve a problem or increase the attractiveness of your home.

2. Do research on careers in real estate. Write a one-page report on the types of real estate personnel and their responsibilities.

3. Obtain a loan application from a bank or other lending institution in your community. Read the application carefully. Describe how the information asked for on the application can help the lending institution decide whether to loan money to an applicant.

11 DESIGNING AND ENGINEERING THE PROJECT

Terms to Know

architectural
  drawings
Construction
  Specification
  Institute (CSI)
consultant
dead load

designing
detail drawing
drawings
electrical plans
elevations
engineering
floor plan

infrastructure
live load
mechanical plans
preliminary designs
presentation model
scale models
section drawing

site plan
specifications
specification writers
structural drawing
structural
  engineering
study model

Objectives

When you have finished reading this chapter, you should be able to do the following:

- Explain the difference between designing and engineering.
- Name two important aspects of design.
- Define live and dead loads and give an example of each.
- Describe the three general categories of engineering.
- Describe the various drawings used in the final design.
- Explain who provides architectural and engineering services.

Every construction project starts with a plan. The plan identifies all the details of the project. It is developed by many different people, such as architects, engineers, drafters, and specification writers.

Creating a design, or designing, is the first step in a construction project. The design of a project is closely related to two other essential steps: project management and building.

The project plan is developed by designing and engineering. **Designing** is the process of deciding what a structure will look like and how it will function. **Engineering** is the process of figuring out how the structure will be built and what structural materials will be used. Fig. 11-1.

The two most important aspects of design are function and appearance. A good design must have a balance between the two. A design that looks nice but does not do what it is supposed to do is not a good design. A design that meets the needs of the owner but does not look good is not a good design, either.

▮▮▮ DESIGNING

Designing a project is a challenge. A design can be entirely new. A design can also result from several ideas combined to meet the needs of a specific project. Everything about the project must be designed: its shape, style, size, and layout.

If a construction project is not well designed, thousands of dollars can be lost. If problems are found in the design after the project has been built, any needed changes may be very costly. Expensive mistakes can be avoided if the design is carefully planned and prepared.

Fig. 11-1. The plans for almost all structures are developed through designing and engineering. This structure is designed to be attractive and functional. It is engineered to house spectators and players safely during sporting events.

Designing a Functional Structure

The first step in designing a functional structure is to gather information about its proposed use. For example, assume that a new school is being designed. The designer must find out how many classrooms will be needed and what size the classrooms should be. The designer also needs to know how many students are expected to attend the school. The number of students affects the size of the walkways as well as the size of the auditorium and the lunchroom. Another thing the designer should know is the approximate age of the students. Very young students may require modified drinking fountains and low classroom cabinets, among other things. Fig. 11-2.

The designer must also be aware of more general design characteristics. These include ease of maintenance, traffic flow, and energy efficiency. These characteristics are common to almost all structures. The design to meet such needs differs according to the type of structure. A good functional design meets the owner's needs, is easy to maintain, has a smooth traffic flow, and is as energy efficient as possible.

Designing an Attractive Structure

The *appearance* of the structure is the way it looks. Appearance is important because it gives an overall impression of the building. The appearance of a structure affects the way people feel about it. It should reflect the activities that take place within the building. For example, a school should not look like a factory. A shopping center that looks like a hospital may not attract many customers. Designers are careful to consider the effect of appearance on the people that use the building.

Fig. 11-2. This school building is designed so it can be used by students who use wheelchairs. Notice how the designer incorporated ramps into the building's design.

The structure must also fit into its surroundings. Fig. 11-3. For example, buildings located in the downtown area of a city usually look different than buildings located in the country. A country home would look as out-of-place in the city as a city home would in the country. Sometimes a structure can fit into its environment even though it is quite different from its surroundings. Examples are the Gateway Arch in St. Louis and the Statue of Liberty in New York. Both are monuments with unique designs, but each fits into its surroundings. Fig. 11-4.

Fig. 11-3. This house was designed to fit into its environment. (*Frank Lloyd Wright, Fallingwater*)

Fig. 11-4. Some structures, such as the Eiffel Tower in Paris, France, fit into their environments even though they are different from their surroundings.

Did You Know?

In designing a building, an architect must consider many different things. One of the things that the architect must consider is scale. In the case of a building, scale is the proportion between one part of a building and its other parts. Proper scale in a building is important if the building is to be attractive.

The temples of ancient Greece have a pleasing scale. Among the many pleasing features of those temples are the columns. In building these temples, the Greek builders employed the base of the column as a building module. A module is a unit of measurement. All of the measurements in the temple were based on this module. This helped ensure a proper sense of scale for the building.

Making Preliminary Designs

The design process begins with the identification of the project. The need for the proposed structure is established. Next, information is gathered and evaluated to see if it is possible and practical to build the proposed structure.

An architect and an engineer are hired to help design the project. They must know the owner's requirements. They must ask some important questions to be able to design the proper building. These questions include:
- For what will the structure be used?
- What type of structure is best suited to the owner's requirements?
- How much will the structure cost?
- What zoning and title restrictions apply?

The opportunity for future expansion and the availability of utilities are also considered at this time.

Once the owner decides that the project is possible and practical, the designer begins his or her work. The designer begins by making **preliminary designs**, or first sketches of what the structure might look like. During this process, many sketches are made. Fig. 11-5. All of the sketches are saved for evaluation at a later time. Not all of these preliminary ideas will be usable. However, some of the ideas shown in the sketches may provide solutions to problems that occur later in the designing process.

The designer seldom comes up with the final design right away. The preliminary designs are usually *refined*, or improved. The best ideas are then selected. Better drawings are made, showing these ideas. Ideas from several different designs may be combined into one new design. The designer then studies and compares the attractiveness, function, and efficiency of each refined design.

Finally, the designer selects his or her best ideas and presents them to the owner. The owner chooses one design to be developed into the final design and makes the decision to go ahead with the project. The plans can now be started.

For Discussion

Suppose that you are preparing the first design sketches for a children's treehouse. What design characteristics would you need to consider?

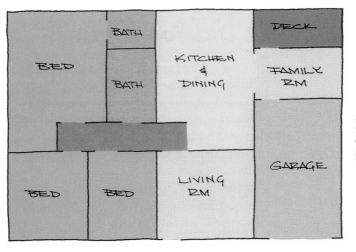

Fig. 11-5. These sketches show some of the designer's preliminary design ideas for a house.

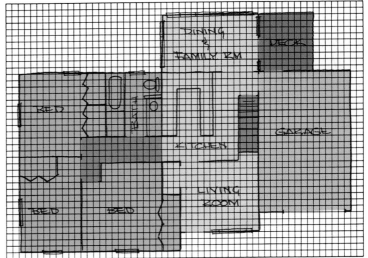

ENGINEERING

The design determines the layout and appearance of the structure. However, other factors must be determined before a structure can be built. Engineers must calculate the sizes and types of materials needed to build the structure properly. They must find the answers to questions such as these:

• What size foundation is needed?

• How much reinforcement is needed in the concrete?

• What size pipes and wires are needed?

• How much air conditioning is needed?

The solutions to these problems can be found by the engineering process.

Engineering a structure means figuring out how it will be put together and how it will work. Fig. 11-6. To know which materials should be used in a structure, engineers must predict how the materials will act under certain conditions.

Fig. 11-6. Before construction of this bridge could begin, engineers had to predict its behavior under high, unfavorable winds.

Structural Engineering

Structural engineering is the process of selecting appropriate materials from which to build a structure. For example, structural engineers use formulas and scientific principles to determine the stress that will be placed on the finished structure. They can then decide which materials to use in a structure and what size each type of material has to be. The foundation and the framework of the walls and roof are the most important structural parts. The structural engineers must be sure that these parts can safely carry the load they are expected to support.

Loads

To engineer a structure, engineers must know how to figure loads. The *load* is the amount of weight the structure will have to support and the forces it will have to withstand. There are two types of loads: dead loads and live loads. The **dead load** of a structure is the combined weight of all its materials. A dead load is constant; it is always there. Concrete, lumber, carpet, and even paint contribute to the dead load.

Their predictions are based on known mathematical and scientific principles.

To engineer the size of a building's foundation, it is necessary to know the type of soil on which the building will be built. The engineer must also know how much weight the soil can carry. The weight of the proposed structure must also be known. Engineers can tell how much a structure will weigh. They can do this by calculating the weights of all the materials to be used. The size of the foundation can be determined from these figures.

Fig. 11-7. Everything you see in this picture is part of the dead load of the structure.

Fig. 11-7. A **live load** is a variable, or changeable load. It is one that is there only some of the time. People and furniture are part of the live load in a building. Snow on the roof and wind pushing against the side of the building are also considered part of the live load. Fig. 11-8.

Safety

After the engineers have done all of the necessary calculations, they study the figures to be sure the building will be safe. When they are satisfied, the engineers add a safety factor into their calculations. A safety factor is extra strength that is built into a structure to provide a wide margin of safety. For example, an elevator designed to hold ten people will hold one or two more before it will fail. It has a safety factor to help prevent accidents.

Fig. 11-8. Snow is part of the live load of a building.

Mechanical and Electrical Engineering

The mechanical and electrical setups of a structure are commonly referred to as its **infrastructure**. Fig. 11-9. For many small projects, the general designer plans the placement and installation of the infrastructure. He or she also specifies the type and size of wires, pipes, and other materials needed to meet safety standards and building codes. Most large projects require the services of mechanical and electrical engineers. These specialists design the complicated mechanical and electrical systems for hotels, large office complexes, and similar structures.

Mechanical Engineering

The mechanical devices in a building are usually classified by system. The plumbing system provides drinking water and carries away wastewater. Every part of this system has to be engineered. The sizes of all the pipes, the hot water heater, and the drinking fountains have to be chosen to fill the needs of the structure. The drain pipes and other parts of the wastewater system also have to be engineered.

Another mechanical system is the HVAC system. *HVAC* stands for *h*eating, *v*entilating, and *a*ir *c*onditioning. The air conditioner, furnace, and ducts must be the correct size.

Electrical Engineering

The installation of electrical wires and devices in a structure must also be carefully engineered. Wire has to be large enough to allow electrical current to flow through safely without overheating the wire. There must be enough electrical circuits to provide power to all motors and lights. Fig. 11-10. By calculating the amount of electrical power needed in the proposed building, engineers can determine correct sizes.

Systems Costs

The best-engineered design is one that does the job efficiently at the least possible cost. It would not make sense to put an expensive, 500-horsepower engine into a small passenger car. It could be done, but the result would not be very efficient. To do so could be an example of *over-engineering*. A 60-horsepower engine

Fig. 11-9. The mechanical and electrical systems are part of a building's infrastructure. Each system is engineered to fill the needs of the structure it will serve.

Fig. 11-10. Electrical systems are engineered to provide the right amount of power for a specific facility. The lighting in a theatre must be specially designed.

could provide enough power for the car and would cost a reasonable amount. Using a 60-horsepower engine is an example of *engineering efficiency*.

Engineering efficiency must also be considered by structural engineers. The engineers want the structure to be safe but not overly expensive. As all the calculations are being made, the engineer looks over all the parts of the design again. He or she checks to be sure that the materials specified are practical and are used as efficiently as possible. Checking the efficiency of materials is often called *value engineering*.

For Discussion

In this section, *dead load* and *live load* have been discussed. The *live load* is described as being variable, or changeable. Examples of live loads include people and furniture. Name some other examples of dead loads and live loads.

THE FINAL DESIGN

After all the calculations have been made, the design becomes final. This final design is used to prepare a set of drawings to be used by the builder to construct the project. Several kinds of drawings are needed to show all the construction information. The **drawings** show the plans for a structure in graphic form. For a large construction project, such as a power plant, there may be more than a hundred drawings in the set. Each of the different types of drawings shows special information that is needed for the project. Symbols are used on the drawings to represent the methods of construction and materials to be used. Fig. 11-11. A set of specifications is also prepared. The **specifications** tell the contractor exactly what materials to use and how to use them. The drawings and specifications give all the information necessary to build the structure exactly as it was designed and engineered. Sometimes a model is also built so that people may see how the finished structure will look.

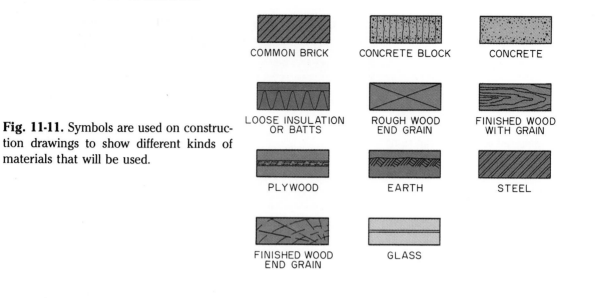

Fig. 11-11. Symbols are used on construction drawings to show different kinds of materials that will be used.

Architectural Drawings

Architectural drawings are those that show the layout of a building. It is impractical to make the drawings the size of the actual structure. Thus, engineers make them *to scale*. This means that actual measurements are converted to smaller measurements that will fit on the drawing sheet. Everything is still in the right proportion, just smaller. For example, one common architectural scale is the quarter-inch scale, in which each ¼ inch represents 1 foot.

Many different kinds of architectural drawings are needed to make every aspect of construction clear. Some of the most common kinds of architectural drawings are floor plans, site plans, elevations, section drawings, and detail drawings.

Floor Plan

The **floor plan** shows the layout of all the rooms on one floor of a building. Fig. 11-12. It also shows the locations of all the walls and other built-in items. A separate floor plan is made for each floor in a building.

Site Plan

The **site plan** shows what the site should look like when the job is finished. Fig. 11-13. The location of the building, new roads and parking lots, and even trees are shown. The finished contour of the earth is also shown.

Elevations

Elevations are drawings that show the outside of the structure. An elevation drawing is made for each side of the structure. The complete set of elevation drawings for a house shows a front, a rear, and two side views. Fig. 11-14. These drawings contain information about the placement of joists, siding, and roofing materials that will be used for the structure.

In addition to exterior elevations, the plans usually include interior elevations. Interior elevations are often needed for the kitchen. The heights of kitchen counters, cabinets, and shelves are shown on an interior elevation.

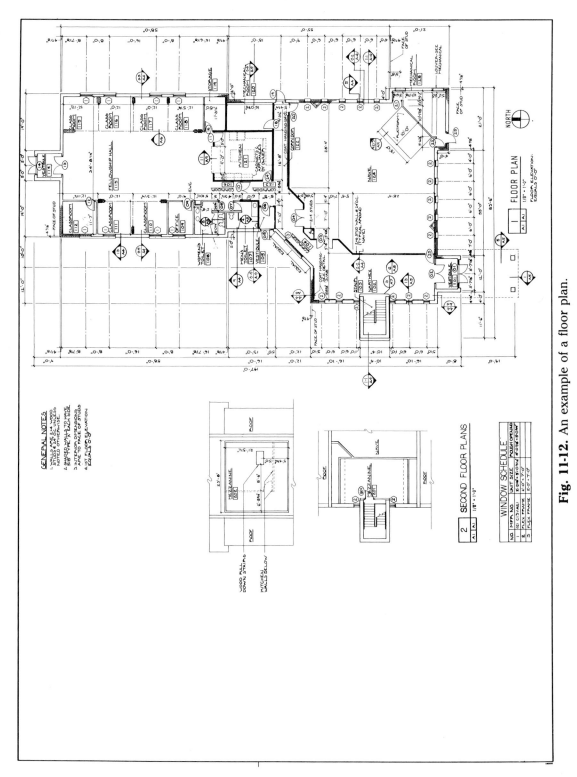

Fig. 11-12. An example of a floor plan.

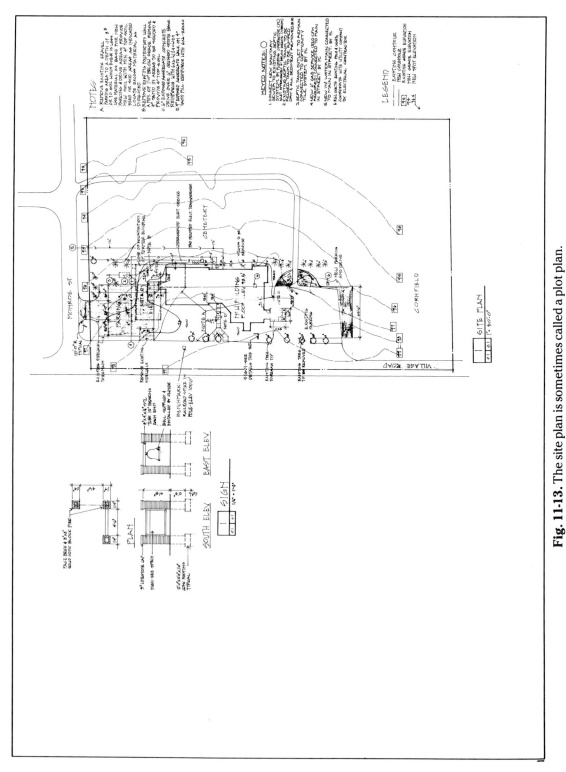

Fig. 11-13. The site plan is sometimes called a plot plan.

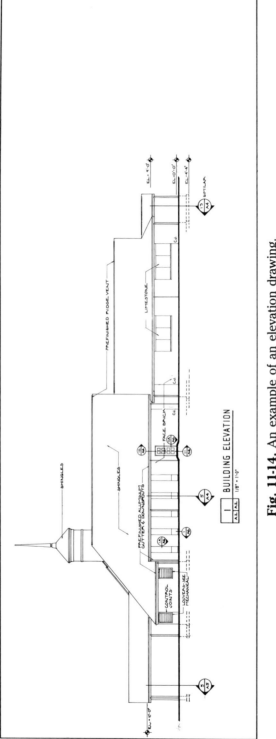

Fig. 11-14. An example of an elevation drawing.

Section Drawings

A **section drawing** is one that shows a section, or slice, of the structure. In this drawing, a part of the building is shown as if it were cut in two and separated. This allows the viewer to see the inside of the structure. Fig. 11-15. If most of the walls will have the same construction, a typical wall section is drawn to show their interior detail. If the walls will have different constructions, a section view is drawn of each kind of wall.

Detail Drawings

A **detail drawing** is one that shows a particular part of the structure. It shows how things fit together. Detail drawings normally are drawn to a larger scale than other drawings. This makes them easy to read and eliminates confusion about each particular point in the structure. Fig. 11-16.

Fig. 11-15. A typical section drawing.

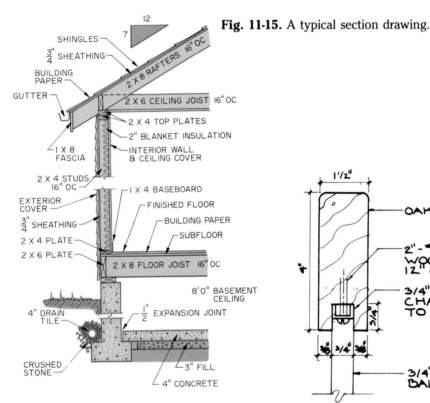

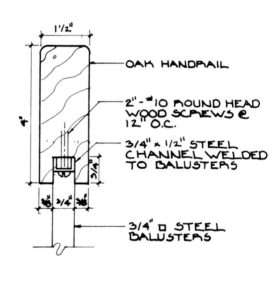

Fig. 11-16. An example of an architectural detail drawing. This is a detail drawing of an oak handrail on a staircase.

5 A	DETAIL
A7 | A7	6"= 1'-0"

Did You Know ?

Architectural drawings were made nearly four thousand years ago by the Babylonians. The Babylonians were a people who lived in what is now Iraq. For example, a stone engraving of the plan of a fortress has been found. Archeologists think that this stone engraving was made about 2000 B.C. The ancient Romans also produced architectural drawings. In about 27 B.C., the Roman architect Vitruvius wrote what might be the first book on engineering drawings.

Structural Drawings

A **structural drawing** gives information about the location and sizes of the structural materials. The size and type of foundations, the location of steel columns and beams, and the placement of roof trusses are found on the structural plan. Fig. 11-17 (page 254). Structural plans are prepared by a structural engineer.

Mechanical and Electrical Plans

The mechanical and electrical plans show how to construct the various systems in a structure. **Mechanical plans** are prepared for the plumbing and piping systems. Fig. 11-18. Mechanical

Fig. 11-18. This plumbing diagram is part of the mechanical plans of a structure.

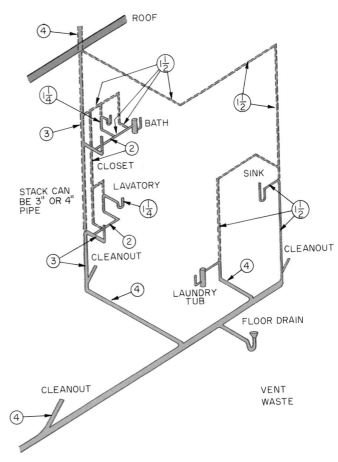

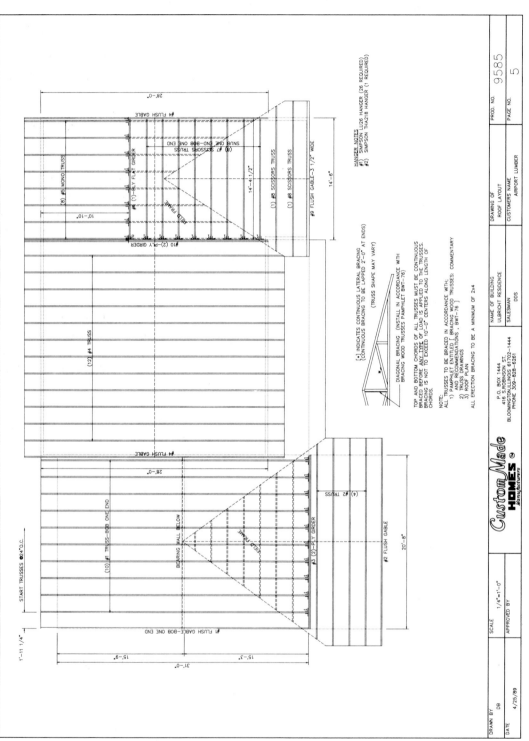

Fig. 11-17. A structural plan shows how the parts of the frame fit together.

drawings are also made to show the construction and placement of the HVAC system. The **electrical plans** show the location of all the light fixtures, switches, and other electrical devices. Fig. 11-19.

Specifications

Drawings alone cannot tell the contractor everything he or she needs to know about a project. For example, it would be difficult to try to describe all the types and colors of finishing materials such as paint, ceramic flooring, and wallpaper. Therefore, written specifications are developed. These describe details that cannot be shown easily on a drawing.

In addition to describing materials, the specifications explain the procedure that should be used to perform each task. Especially difficult or unusual procedures are described in detail. Construction standards are also included to make sure the finished project will conform to federal, state, and local requirements.

Writing the Specifications

The specifications for a major construction project can be so extensive that they are bound in two or three volumes. To make sure that all the information can be found easily, the specification writer must keep it in a logical order. Most specifications for large construction projects follow the format presented by the **Construction Specification Institute (CSI)**. The CSI guidelines are presented in outline form. Fig. 11-20. Every type of construction job is classified in one of 16 categories. The categories are meant to follow the actual order of construction as far as possible. They also keep all the information about a subject in one place. For example, all the instructions for plumbing installation would be found together in one part of the specifications.

Writing the specifications for a large project is an involved and difficult procedure. The specifications must meet the requirements of the owner, the architect, and the engineer. They must also comply with the rules and regulations set forth by city, county, state, and

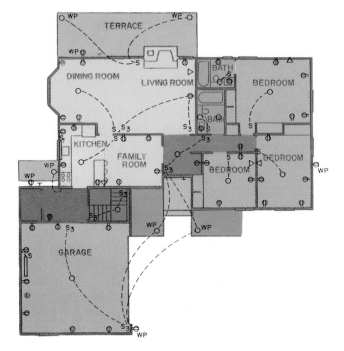

Fig. 11-19. An example of an electrical plan.

○	CEILING FIXTURE
-○	WALL FIXTURE
Ⓓ	DROP CORD FIXTURE
○$_{WP}$	WEATHERPROOF FIXTURE
▭	FLUORESCENT FIXTURE
S	SINGLE POLE SWITCH
S$_3$	3-WAY SWITCH
⊣	DOOR CHIME
⊕	DUPLEX OUTLET
⊕$_{WP}$	WEATHERPROOF OUTLET
⊕	HEAVY DUTY OUTLET
◁	TELEPHONE

2.02 Chalkboards and Tackboards

 .1 Types

   a) Where chalkboard (CB) units are indicated, they shall be Claridge 4' high factory-built units, Series 4, with continuous chalk trough, flat type on lower edge of chalkboard. Provide individual units as indicated.

   b) Where "full height" units are indicated, they shall consist of 6' high, 4' wide units with color-matched "H" trim at all interior vertical joints. Perimeter trim shall have same appearance as Series 4 factory-built units except without continuous chalk trough on lower edge. Provide two (2) 2'-0" pieces of magnetic chalk trough at each teaching area.

 .2 Chalkboard Construction: Shall be Claridge porcelain enamel steel with the following characteristics:

   — 24 gauge "Vitracite" face.
   — 3/8" particleboard or "Duracore".
   — .005" aluminum sheet panel backing.
   — Total thickness of 1/2".
   — Warranted for the life of the building.
   — Provide chalk trough for each unit.
   — Provide map rail for each unit.
   — Chalkboard color shall be as selected by Architect from not less than six (6) standard colors.

Fig. 11-20. The CSI format for specifications is used for most large construction projects. This set of specifications is for a new school building. Notice the detailed specifications for the chalkboard.

federal governments. While meeting all of the above requirements, the specifications must remain within the owner's budget.

A great deal of knowledge and experience is needed to put the specifications together. Most large construction companies have specialized employees called **specification writers** to do this work. Specification writers must have a good understanding of construction practices and an up-to-date knowledge of materials. They must also be familiar with the current costs of materials and with the regulations that apply in their area.

Using the Specifications

Specifications are important in every phase of the project. They are sent to several contractors before a contractor is chosen for the job. The contractors use the specifications to estimate the cost of materials and labor. They use that information to bid on the contract. During construction, the chosen contractor must follow all the guidelines set forth in the specifications. If the drawings and the specifications disagree, the contractor follows the specifications. When the project has been completed, inspectors check the structure against the specifications. The specifications help them make sure that the work was done correctly and that the proper materials were used.

Models

Scale models are sometimes made to help visualize how the final design will look. These models have proportions that the finished building will have. However, they are built on a much smaller scale. There are two basic types: presentation models and study models.

Presentation Models

A model that is made to show people how the finished design will look is called a **presentation model**. Fig. 11-21. Some people have difficulty imagining the finished appearance of the building by just looking at the plans. A three-dimensional model helps them understand the structure. Models of many commercial projects are put on public display. Perhaps you have seen a presentation model of a bank, church, or other structure.

Study Models

Sometimes a **study model** of a design is made so that the design can be tested. For example, a scale model of a bridge can be loaded with weights and observed. A scale model of a building can be tested in a wind tunnel to see how it might act in a hurricane.

Today most study models are made by computer simulation. All the mathematical information about a structure is entered into a computer. The model appears on the screen. Fig. 11-22. The engineer or architect can change the parts of the model or turn it around. Then a simulation program is run that *simulates*, or imitates, what would happen to the structure under certain conditions. The information can then be printed out. It can then be studied to find weaknesses and ways to improve the building.

Fig. 11-21. A presentation model helps people visualize the final design.

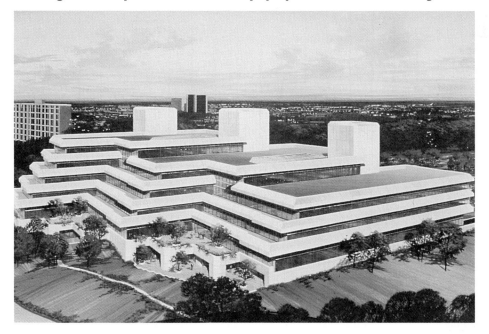

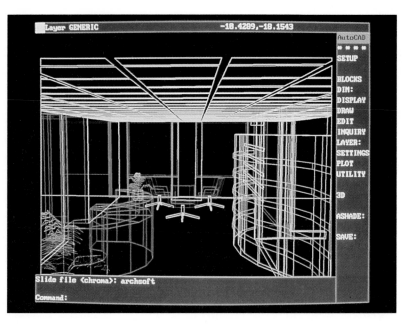

Fig. 11-22. A computerized study model can be tested using a simulation program. In this study model the size of the roof and wall beams can be changed.

HEALTH & SAFETY

Some bridges have collapsed during heavy storms. To help prevent such disasters, engineers have various techniques available. One of these involves the use of a wind tunnel. In a wind tunnel, fans are used to create a stream of air. The force and direction of this stream of air can be controlled. This controlled stream of air can then be directed on a scale model.

For example, a scale model of a bridge might be placed in the wind tunnel. Then the stream of air might be adjusted to simulate the force of a hurricane. This will allow engineers to closely observe the effects of high winds on the scale model. By studying the results, engineers can plan improvements that will help make the bridge safer.

For Discussion

This section discusses the various plans needed in the building of a modern house. Clearly, houses built today are different from those built one hundred and fifty years ago. What would be missing from the plans of a house one hundred and fifty years ago that would be present in the plans for a house today?

ARCHITECTURAL AND ENGINEERING SERVICES

The architect and engineer often work together to design a project. They agree to work for the owner for a fee. The amount of the fee depends upon the size of the project. A company that does designing and engineering is known as an A/E (architectural and engineering) firm.

Just as a medical doctor must have a license to practice medicine, architects and engineers must have a license to practice their professions.

Fig. 11-23. The architect's or engineer's seal and signature on a set of drawings means that he or she is responsible for the design.

The license means that the person is legally qualified to design and engineer a structure. In fact, he or she must sign the drawings and stamp his or her official seal over the signature. Fig. 11-23. This shows that the person is legally responsible for the designing and engineering work on the drawing.

Architects

The main job of the architect is to design and create plans for structures. Architects design houses, churches, office buildings, and shopping centers. Fig. 11-24. They also design community projects such as auditoriums and sports arenas.

The architect is the supervising designer in most building projects. During project construction, the architect checks that the contractor is following specifications. The architect approves samples of the materials to be used on the project. He or she must also approve substitute materials or changes in construction methods. The architect works for the owner and with the contractor to make sure that construction is done according to the plans.

Fig. 11-24. An architect designed the layout and appearance of this shopping mall.

Fig. 11-25. Each of these construction projects — the water tower and the bridge — was designed and engineered by an engineer.

Engineers

Engineers help design and engineer all types of structures. However, they are usually the supervising designers of heavy construction projects such as bridges, roads, and utility systems. Fig. 11-25. The engineer's duties and responsibilities are similar to those of the architect. The engineer makes sure that the specifications are being followed and that the construction is done correctly. He or she also inspects the project materials and workmanship.

Did You Know?

The earliest engineers were military engineers. They were responsible for the design and construction of many of the surviving fortifications of the ancient world. In Asia, one example of the work of military engineers is the Great Wall of China, which extends 1,500 miles.

In Europe, examples of the work of Roman military engineers can still be seen. Among them is Hadrian's Wall, which extended for 73 miles. It was intended to protect the northern frontier of Roman Britain. The term civil engineer was not used until the eighteenth century. It was used to indicate that the civil engineer was concerned with the building of public works.

Consultants

A **consultant** is an expert in a specific area. He or she may be called upon to give advice on a certain part of the design of the construction project. The consultant works as a subcontractor for part of the designing and engineering work.

It is common practice for an architect to hire consultants for special parts of the design work. For example, an architect may be designing a new concert hall. The design of the hall will have an effect on the sound the audience will hear when music is played. The architect may call in a consulting engineer who specializes in *acoustics*, or sound. The consultant provides advice about the overall design as it applies to the acoustics. Structural, foundation, and mechanical engineers often serve as consultants. Another commonly used consultant is an interior designer. He or she works with the architect to design the appearance of the interior of the building. Fig. 11-26.

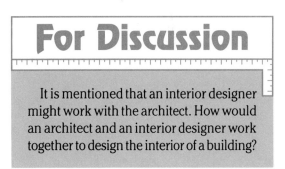

For Discussion

It is mentioned that an interior designer might work with the architect. How would an architect and an interior designer work together to design the interior of a building?

Fig. 11-26. An interior designer helped the architect design this office.

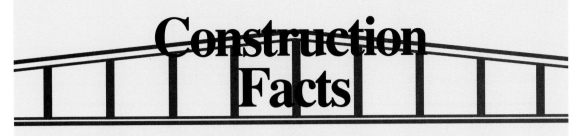

Construction Facts

JOHN ROEBLING AND THE BROOKLYN BRIDGE

In the late 1860s, the people of New York decided that a bridge was needed to connect Brooklyn and Manhattan. They called upon an engineer who was considered the lead-

ing bridge designer of that time: John Roebling. Mr. Roebling suggested a new design for the bridge. He had developed a new cable-weaving technique that would enable him to design a strong, safe bridge that could hang from wire cables.

According to his design, four fifteen-inch-thick cables would hold up the bridge. Each cable would be made up of 19 strands of wire designed by Roebling. Each strand would be made up of 278 separate wires. Each wire would be continuous—there would be no wire ends to break apart.

The people approved. They put John Roebling in charge of building the new bridge. Unfortunately, John Roebling died before the construction could begin. His son, also a civil engineer, took charge of the project. Washington Roebling began the construction in January, 1870, and completed the project according to his father's design in 1883.

John Roebling designed and engineered the bridge carefully. His design far exceeded the safety limits needed for traffic of the 1800s. As a matter of fact, the safety factor he used is adequate for today's heavy traffic flow. The Brooklyn Bridge is still considered a mighty engineering feat. The enormous weight of the massive road is still supported effortlessly by the weblike structures of wire.

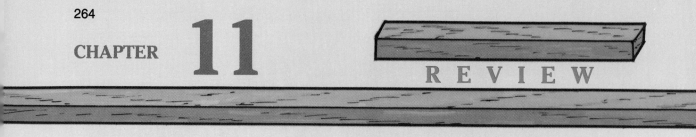

Chapter Summary

Every construction project requires a project plan. A project plan requires designing and engineering. A structure must be functional and attractive. In designing a project, preliminary designs are made. Engineers figure out how the structure will be built. They select the appropriate materials. Engineers also figure the loads and determine whether the building will be safe. The mechanical and electrical setups of a building are referred to as its infrastructure. The design of the infrastructure is the work of mechanical engineers and electrical engineers. The drawings for a building show the plans for a structure in graphic form.

The specifications tell the contractor what materials to use. Architectural drawings show the layout of the building. Architectural drawings include floor plans, site plans, elevations, section drawings, and detail drawings. Structural drawings give information about the location and sizes of structural materials. Mechanical and electrical plans show how to construct the various systems in a structure. Specifications for a construction project must be accurately written. Models are sometimes used to help visualize how the final design will look. Architects and engineers often work together in designing a project.

Test Your Knowledge

1. What is the difference between designing and engineering?
2. What are two major factors that need to be taken into consideration in the design process?
3. What is the difference between a live and dead load? Give two examples of each.
4. Why do engineers include a safety factor in their calculations?
5. Name five types of architectural drawings.
6. What are specifications? What purpose do they serve?
7. What two types of models are used for construction purposes?
8. How can a computer simulation be used to test a proposed design for a structure?
9. What kinds of projects do architects usually design and engineer?
10. What kinds of projects do engineers usually design and engineer?
11. Why are consultants used to help design some projects?

REVIEW

Activities

1. Sketch a floor plan of your house or apartment. Analyze the layout. How functional do you think it is? Write a description of your opinion.

2. Look around your community. Identify by name or address two structures that you think are well designed and one that you think is poorly designed. Give reasons for your opinions.

3. Choose any room in a building. Make a list of all the things you see. Tell whether each item contributes to the live load or dead load of the building.

4. Obtain a set of plans for a construction project. Study the drawings and see if you can find examples of the different types of drawings.

5. Make a scale presentation model of your house or apartment.

12

CONSTRUCTION PROCESSES

Terms to Know

armored cable
batter boards
bearing-wall
  structure
conduit
excavating
finishing stage
floating

footing
forms
foundation
foundation wall
frame structure
laying out
load-bearing ability

molding
nonmetallic
  sheathed cable
roof truss
roughing in
screeding
service drop

service panel
sheathing
shoring
site
slump test
superstructure
troweling

Objectives

When you have completed reading this chapter, you should be able to do the following:

- Explain the process of preparing a construction site.
- Describe three major types of structural work.
- Explain the difference between a frame structure and a bearing-wall structure.
- Describe three basic types of utility systems.
- Describe five major types of finish work.

Construction processes are all the procedures that are used to build a construction project. Most construction projects are very complex. They require a great many processes before they are ready to be used. For a building, the construction processes include activities ranging from the initial preparation of the site to the installation of the wallpaper. Most of these processes are carried out by workers who specialize in just a few processes. These specialized workers are able to work quickly and efficiently.

In reading this chapter you will learn about many of the major processes required to complete a structure. As you read about these processes, you will see their complexity. This will help you understand why construction workers must specialize in one or only a few processes.

PREPARING THE SITE

The land on which a project will be constructed is called the **site**. Fig. 12-1. The plans for a project give the details about how the site should look when the project is complete. These plans also tell the workers what needs to be done to the site before the building is started. Plans were discussed in Chapter 11.

Three major steps are involved in preparing a site for construction. First, the land is surveyed to establish its boundaries. Then the site is cleared. The last step is to lay out the boundaries of the building.

Fig. 12-1. A site is the land on which a construction project is built.

Surveying the Site

The first step in preparing the site is to identify the boundaries of the property. Surveyors measure the property to establish property lines. They use a transit and a measuring tape called a *chain* to measure very accurately. Fig. 12-2. The slightest mistake in measurement can cause the property lines to be wrong. After measuring the land, the surveyors set stakes at each corner of the property.

Clearing the Site

Once the property boundaries have been defined, the next step is to remove anything that is in the way of the new construction. Trees, dirt, and old buildings are a few of the things that may need to be removed. Bulldozers are usually used to push trees, excess dirt, and other unwanted materials out of the way.

Demolition, or wrecking, is used to clear buildings from a site. There are various ways of demolishing old structures. One common demolition method is to use a crane with a wrecking ball. Other demolition methods include using bulldozers or explosives to break up a structure. Fig. 12-3.

Fig. 12-2. This surveyor is using a transit to establish the boundaries of a site.

Did You Know?

Explosives can be used to demolish tall buildings. Some demolition companies specialize in this type of site preparation for construction. The explosives are precisely placed. Also, the amount of the explosive varies from one location to the next in the building that is to be demolished. Used in these ways, the explosives can be used to demolish a building without damaging nearby buildings. This method of demolition is sometimes used to remove fairly tall office buildings.

Laying Out the Site

Laying out the site is the process of identifying the location of the proposed structure on the property. The corners and edges of the building to be constructed are marked with stakes and string. Proposed parking lots and roadways are also marked at this time.

To lay out a site, construction workers use the boundary stakes left by the surveyors as points of reference. They measure from these stakes to the corners and edges of a proposed building.

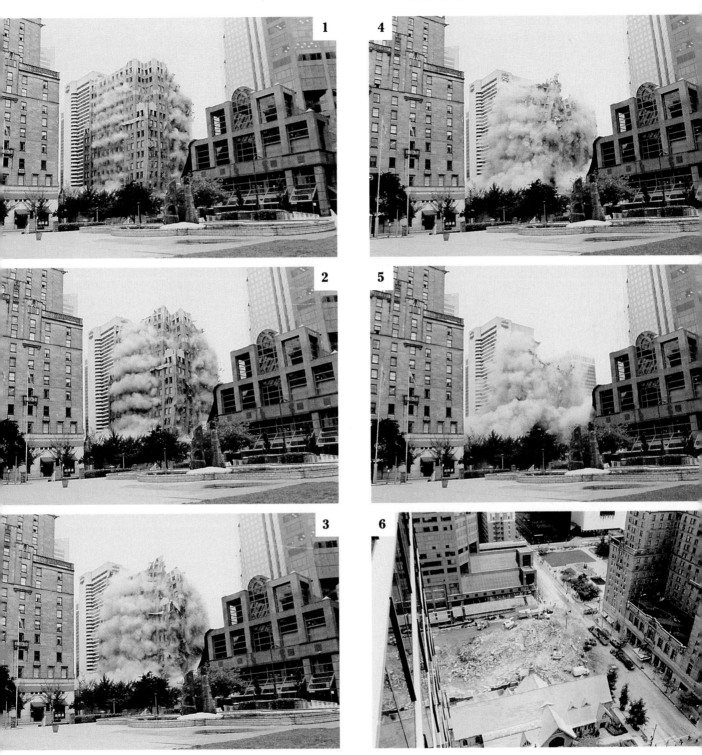

Fig. 12-3. Explosive demolition is a fast way to demolish an old structure. This series of photos shows the stages in a building's collapse.

The boundaries of the building are usually marked by **batter boards**. A batter board is a board held horizontally by stakes driven into the ground. String is used to connect the batter board with another at the opposite end of a wall. The batter boards are placed outside the building's boundaries so that the attached strings cross over the corners of the building. Fig. 12-4. This ensures that workers will know where the boundaries are even after excavation begins and the boundary stakes are removed.

The building layout is important because it shows the shape and size of the planned structure. Like the surveyors, construction workers must measure very accurately. If there are any mistakes in measuring at this point, the whole structure will be built in the wrong place or to the wrong size.

Excavating and Earthworking

The next step in constructing a building is to condition the earth. **Excavating**, or digging, can begin as soon as the site has been laid out. In addition to excavating, other earthworking—such as leveling, grading, and soil stabilization—may be needed. Soil may also need to be removed from or added to the site so that the proper elevation is achieved.

Excavating

One of the main reasons for excavation is to provide a place for the foundation, or base, of the structure. The foundation is an important part of the structure because it supports the structure's entire weight. The foundation must be strong and rigid enough to hold up all the building materials used in the structure. It must also support the weight of the furnishings and the people who will use the finished structure.

The type of excavation depends on the project and its location. Sometimes the hole must be dug straight down. For example, the foundation for a new skyscraper being built between two existing buildings would need a deep, narrow hole. Fig. 12-5. The walls of the excavation must be dug carefully so that the sides do not cave in. This kind of excavation is called *trimming and shaping*.

Another kind of excavation must be done to make trenches for pipelines. This type of excavation is called *trenching*. Fig. 12-6. Water, sewer, or natural-gas pipelines are placed in the trenches. Then the earth that was dug from the trench is replaced.

Leveling and Grading

Leveling and grading are processes that change land elevation and slope. They do this by filling in low spots and shaving off high spots.

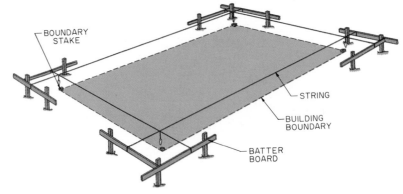

Fig. 12-4. Batter boards and string can be used to lay out a building's location.

Fig. 12-5. Excavating in a confined area is a difficult job. It must be done carefully so that the surrounding structures will not be damaged. This building will be demolished. Then excavation will begin for the foundation of a new six-story office building.

This is especially important in building roads and parking lots. Fig. 12-7. The proper elevation and slope are indicated by the surveyor. Workers follow the surveyor's guidelines to level and grade the site.

Stabilizing the Soil

Before a structure can be erected on a site, the soil must be firm and stable. The soil must not move or shift beneath the foundation. The first step in soil stabilization is to determine the **load-bearing ability** of the soil. This is the amount of weight that soil can safely support without shifting. The load-bearing ability of the soil is measured by a geologist or soil engineer.

If the soil is too weak to support the proposed structure, it must be stabilized. One of several methods may be used. One method of stabilizing

Fig. 12-6. This trenching machine is being used to dig a trench for an underground drainage pipe.

Fig. 12-7. This scraper is being used to level the earth for a road project.

soil involves pounding it with heavy rams or rolling it with heavy rollers. These methods make the soil more compact. It can then bear heavier loads than would otherwise be possible.

When soil must be stabilized around a hole or trench, sheathing and shoring are used. The banks or walls of soil are covered with sheets of steel called **sheathing**. The sheathing is usually held in place by long strips of metal driven into the earth. This method of holding sheathing in place is called **shoring**. Shoring can also be used without sheathing to stabilize soil and reduce the chance of cave-ins. Fig. 12-8.

Soil can also be made firm by adding certain chemicals to it. Loose soil can be filled with a chemical solution that reacts with the soil to make it harder. The hardened soil can then provide a solid base for construction work.

For Discussion

The text discusses the various ways in which a site is cleared. Is there any construction taking place near your school? What was done to clear the construction site?

Fig. 12-8. This metal shoring prevents the soil from caving in.

║║║ BUILDING ║║║ THE FOUNDATION

When the site work has been completed, work on the structure begins. The structure is made of two basic parts: the foundation and the super-structure. The **foundation** is the part of the structure that is beneath the first floor. It includes the footing and the foundation walls. The rest of the building, beginning with the first floor, is called the **superstructure**. You should note that some foundations extend above the ground. Therefore, the generalization that everything below ground level is the foundation and every-thing above is the superstructure is incorrect. You will learn more about superstructures later in this chapter.

The **footing** is the part of the structure that distributes the structure's weight. It is usually made of reinforced concrete. The footing must be placed beneath the *frost line*. It must be deep enough so that the soil around it will not freeze in winter. Soil above the frost line expands as it freezes. Therefore, footings placed above the frost line are unstable.

The wall that is built directly on the footing is called the **foundation wall**. It transmits the weight of the superstructure to the footing. Fig. 12-9. The foundation walls for most structures are made of steel-reinforced concrete.

Preparing for the Foundation

Before the concrete can be poured for a foun-dation, two major preparatory steps must be completed. First, forms must be assembled to contain the concrete. **Forms** are the molds that

Fig. 12-9. The foundation wall transmits the weight of the superstructure to the footing, which distrib-utes the weight to the stable soil around it.

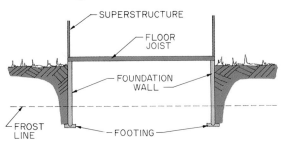

contain the concrete until it hardens. They are made in the shape that the finished concrete should have. Fig. 12-10. Forms can be either custom-built or patented. *Custom-built forms* are wooden forms made by carpenters on the site. This kind of form is used only once. *Patented forms* are made in a factory by a company that then patents the design. These reusable forms are bolted together in sections. After the concrete has set, the forms are disassembled and stored for the next job.

When the concrete forms have been securely fastened in place, steel reinforcement is prepared and put into the forms. The reinforcement is a very important part of the foundation. It pro-

Fig. 12-10. Concrete assumes the shape of the form into which it is placed.

vides extra strength to keep the concrete from breaking under the weight of the structure. Fig. 12-11.

The size and placement of the steel reinforcement are calculated to give maximum strength. Depending on the shape and thickness of the footing, either wire mesh or reinforcing bars are used. There may be several different layers of reinforcing in a slab of concrete. All the reinforcing is connected to make a rigid framework that will stay in place.

Pouring the Concrete

Once the forms and reinforcement are in place, the concrete can be poured. The concrete is usually delivered to the site in cement trucks. These trucks thoroughly mix the concrete as they are driven to the job site. Concrete from these trucks is sometimes called *ready-mix*. This is because the concrete is already mixed and ready to pour into the forms.

Before the concrete is poured into the forms, two types of samples are taken for testing. A

Fig. 12-11. Steel reinforcement gives the concrete extra strength.

slump test is made to check the workability of the concrete. Fig. 12-12. Another kind of sample is taken in special test cylinders. This sample is later tested in a laboratory to determine the finished strength of the concrete.

The concrete is poured in the forms using one of several methods. It can be poured directly from the truck by means of a chute. It can also be pumped into the forms using a concrete pump and hose. Another common way to pour concrete is to use a bucket on a crane. The bucket is filled with concrete. The crane lifts and positions the bucket over the form. Then the bucket is emptied and the concrete flows into the form.

As the concrete is poured into the form, special care is taken to make sure all parts of the form are filled. A mechanical vibrator is used to help remove air pockets and to make sure that the concrete fills the form thoroughly. Next the concrete is screeded. **Screeding** is the process of moving a straight board back and forth across the top of the form. This removes any excess concrete and levels the top of the concrete.

Concrete slabs such as floors and sidewalks must be smooth and level. For these surfaces additional finishing steps are required. After the concrete is screeded, the surface is floated. **Floating** is the process of moving coarse aggregate down into the concrete, leaving only fine aggregate and sand on top. This is accomplished by moving a wooden or magnesium *float* back and forth over the surface. Then, after the concrete has begun to set, a steel trowel is used to smooth the surface. This final smoothing is called **troweling**. Fig. 12-13.

Fig. 12-13. A power trowel is used to finish concrete.

Fig. 12-12. This worker is doing a slump test on concrete. Basically, this test involves placing the concrete in the mold shown here. The mold is then removed and the slump of the concrete is measured. This is the distance from the top of the mold to the top of the concrete sample.

After the surface of the concrete has been finished, the concrete is allowed to cure. A chemical reaction takes place between the cement and the water to give concrete its strength and hardness. For proper curing, however, the concrete must not dry too fast. To keep the moisture in, the concrete is covered with plastic or sprayed with a liquid sealer. Fig. 12-14. After the concrete is sufficiently cured, the forms can be removed.

HEALTH & SAFETY

For reasons of health and safety, many new construction materials have been introduced. In past years, lead-based paint was widely used in interior decoration. The health hazards of lead are now widely realized. Now, water-based paints are in wide use. In the past, plumbing systems were constructed using pipe that contained a high percentage of lead. Today, pipes of other materials, including types of plastics, are in general use.

For Discussion

You have probably seen large concrete trucks on a construction job. These trucks carry the concrete in a large drum that turns on the back of the truck. What are the obvious advantages of having concrete delivered to a construction site in that way? Can you suggest some ways in which preparing and pouring concrete would have been different in the days before such trucks?

BUILDING THE SUPERSTRUCTURE

The superstructure is built on the finished foundation. The superstructure includes all of the structural parts above the foundation of a

Fig. 12-14. Spraying a liquid sealer onto new concrete seals in the moisture so that the concrete does not dry too fast.

building. It also includes the roofing materials, windows and doors, and all of the finishing materials that are used.

The three main structural parts of a building are the floors, walls, and roof structure. The floors divide the building into levels. The walls divide it into rooms. The roof structure provides support for the cover over the building.

Floors are made of wood or steel frames or of solid concrete. In a house, the ground floor may be made of concrete or of a wood frame covered with wood. The second floor usually consists of a wood-covered wood frame. In a commercial building, the ground floor is usually a concrete slab. Other floors have steel or concrete frames. These floor frames are then covered with a concrete surface.

The type of walls used for a building depends on the type of structural support the building has. You will learn more about structural support later in this chapter. In general, walls can be made of wood, reinforced concrete, steel, or masonry.

Most buildings have a framed roof. The frame spans the distance from wall to wall. Roof trusses are frequently used because they are both strong and lightweight. Fig. 12-15. A **roof truss** is a preassembled frame of wood or steel that is designed to support a roof. Roof trusses are lighter than other types of roof supports, but they are just as strong.

Structural Support

Structures are commonly classified according to how the weight of the structure is supported. There are basically two types of structures. A **bearing-wall structure** is one in which heavy walls support the weight of the building. This type of structure has no frame. The outside walls are usually *bearing walls*; that is, they are weight-supporting walls. In addition, some of the inner walls are bearing walls. Bearing walls are usually made of concrete blocks or of solid concrete. Fig. 12-16. This type of construction is most often used for low buildings of one or two stories.

A **frame structure** is one in which a frame supports the weight of the building. The frame of a frame structure is made up of many connected frame members that are covered by sheathing. The frame members carry the weight of the structure and its contents. The framing can be made of wood, steel, or reinforced concrete.

Fig. 12-15. Roof trusses are lightweight, yet strong enough to bear the weight of roofing materials.

Fig. 12-16. Concrete is often used for bearing walls.

Wood framing is used for many houses and other small structures. Carpenters build and erect the wood frame at the site. Fig. 12-17. Large commercial structures, such as office buildings, have frames of steel. Fig. 12-18. The steel for these frames is prepared in a fabricating shop. Ironworkers assemble and erect the steel on the site according to the building plan. The steel parts are bolted, riveted, or welded to make a rigid frame.

Other large structures are supported by frames of steel-reinforced concrete. The concrete can either be poured at the site or preformed elsewhere. Preformed concrete frames are assembled at the site in much the same way that steel frames are assembled. Fig. 12-19.

Fig. 12-17. Parts of a typical wood-frame wall.

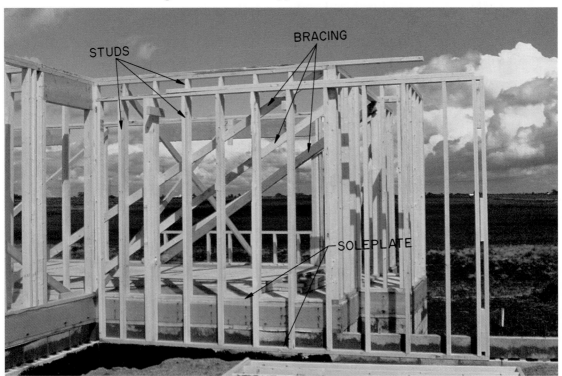

Fig. 12-18. This hospital is a steel-frame structure.

Fig. 12-19. Preformed, steel-reinforced concrete framing members are assembled at the construction site.

Enclosing the Structure

Before any further work can be done to the inside of the building, it must be enclosed to protect it from the weather. Exterior walls must be finished. Windows and doors must be installed, and the roofing must be completed. The appropriate materials for enclosing the structure are usually chosen by the architect.

Exterior Walls

Exterior walls perform many functions in a structure. They protect the inside from the weather and provide privacy for the occupants. They can also be decorative, providing a pleasing view from the outside.

Fig. 12-20. The sheathing on this house will help protect it from the weather.

The first step in completing the exterior walls is to apply the sheathing. **Sheathing** is a layer of material that is placed between the framing and the finished exterior to provide additional insulation. Fig. 12-20. Sheathing usually consists of sheets of plywood, fiberboard, or plastic foam panels that have been nailed or screwed to the framework.

After the sheathing has been installed, the exterior walls are completed using various materials. Fig. 12-21. The following list describes a few of the most common possibilities.

- *Panels* are large sheets of exterior finishing material made of wood, vinyl, aluminum, steel, or marble. They are nailed, bolted, or glued to the framework of buildings.
- *Siding* consists of long, narrow strips of wood, aluminum, steel, or vinyl. It is applied over sheathing and is nailed to the studs of a wood frame. Siding is usually used on residential structures.
- *Masonry* is one of the most practical coverings for exterior walls. Masonry walls include brick, concrete block, and stone. These walls require little maintenance and are fireproof.

Windows and Doors

Wall openings are enclosed with windows and doors. Windows provide light and ventilation. Doors provide a method of entry and exit from the building. The windows and doors are held in place by frames of wood, aluminum, or steel. They are fastened according to the building plans.

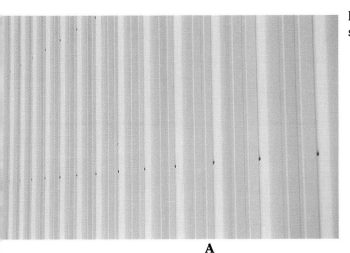

A

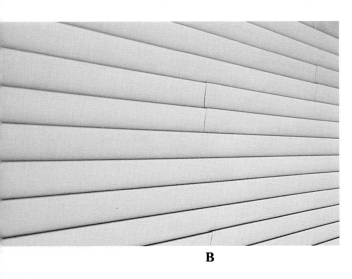

B

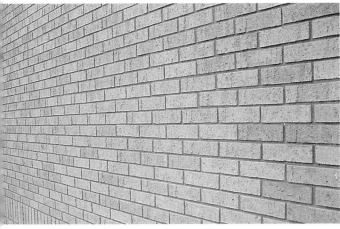

C

Fig. 12-21. Types of exterior walls: (a) panels; (b) siding; (c) masonry.

Roofing

Several kinds of roofing materials are available to cover the roof framing and sheathing. Each is adapted to a certain kind of roof. The following are a few of the most common roofing materials.

Shingles are thin, flat pieces of asbestos, wood, asphalt, or fiberglass. Shingles are applied over a layer of roofing felt, which covers the sheathing. Fig. 12-22. The shingles are overlapped to keep water out. They are suitable for use only on sloping roofs. When properly applied, they are attractive, waterproof, and long lasting.

Built-up roofing is made by alternating several layers of roofing felt and hot bitumen. Bitumen is liquid asphalt or coal tar pitch. When the bitumen cools, it forms a waterproof seal. After several layers of felt and bitumen have been built

Fig. 12-22. Shingles are used on many sloping roofs.

up, a layer of gravel is put on top. The gravel reflects the sun and helps protect the roof from severe weather. Built-up roofing is used mainly on flat roofs. Fig. 12-23.

Sheet metal roofs may be found on industrial, commercial, and farm buildings. Fig. 12-24. They are made of metal sheets that can be attached directly to a metal or wooden frame. No sheathing is needed between the frame and the metal sheets. The metals most commonly used are copper, corrugated steel, and aluminum. Sheet metal is used only for sloping roofs.

Membrane roofing is a relatively new kind of roofing that can be used on flat roofs. Thin rubber or plastic sheets are stretched over the roof to form a continuous membrane. The edges of the membrane are fastened to the structure with an adhesive. Fig. 12-25. The membrane stretches and gives with the slightest movement of the roof. The plastic or rubber is resistant to heavy rain, the damaging ultraviolet rays of sunlight, and heat. The sheets are made at a factory to the correct size and shape. Then the roof is folded and shipped to the site. There it is unfolded and fitted over the roof.

Fig. 12-24. Sheet metal is a practical type of roofing that is long lasting and requires little maintenance.

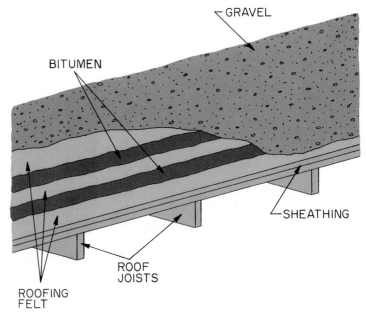

Fig. 12-23. A built-up roof is used on flat roofs.

GRAVEL

BITUMEN

SHEATHING

ROOF JOISTS

ROOFING FELT

Fig. 12-25. Some flat-roofed buildings have a membrane roof. The membrane stretches from one side of the roof to the other to form a waterproof seal.

For Discussion

As mentioned, the frame structure supports the weight of the building. Wood frames, steel frames, and steel-reinforced concrete frames are discussed. The type of frame used will depend on the type of building. Is there any construction around your school? If there is, what type of frames are being built?

INSTALLING UTILITIES

After the frame of the building has been enclosed, the utilities are installed. The three basic types of utilities are plumbing systems; electrical systems; and heating, ventilating, and air conditioning (HVAC) systems. Utility installation is a two-stage process. The first stage, **roughing in**, is the installation of the basic pipes and wiring that must be placed within the walls, floor, and roof. This must be done before the interior walls are covered. Fig. 12-26.

Later, after the walls have been covered, the installation of the utilities is finished. The **finishing stage** of utility installation readies the utilities for use. Faucets and other plumbing fixtures are installed. Light switches and fixtures, wall receptacles, and plates for telephone and cable television connectors are also installed at this time.

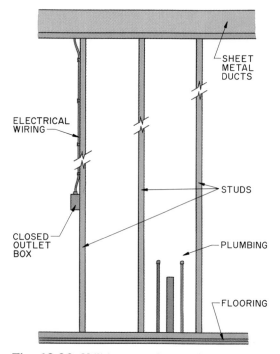

SHEET METAL DUCTS

ELECTRICAL WIRING

STUDS

CLOSED OUTLET BOX

PLUMBING

FLOORING

Fig. 12-26. Utilities must be roughed in before interior walls can be finished.

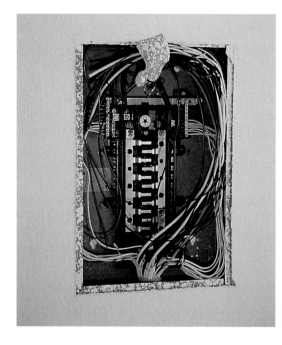

Electrical Systems

Electrical utilities provide power for lights, appliances, heating and cooling systems, and communication systems. Electrical utilities are installed by electricians. At the roughing-in stage, the electrician marks the locations of all the switches, junction boxes, and fixtures. He or she drills holes so that the wires can be pulled through walls and floors. Then the electrical boxes and wires are installed. Fig. 12-27. The wires are pulled into the boxes and left there to be connected at a later time. At this point, roughing in is complete. An electrical inspector checks to make sure that the wiring is acceptable and has been installed according to building codes.

Circuitry

Electrical power enters a building through a service drop. The **service drop** is the wiring that connects a building to the electric company's overhead or underground wires. Fig. 12-28. A meter located near the service drop measures the amount of power used. Near the meter, but inside the building, a service panel is installed. The **service panel** is a box that contains circuit breakers for each individual, or *branch* circuit. It also contains a *main* circuit breaker. This controls all of the individual circuits collectively.

Each branch circuit provides electricity to the outlets, lights, and switches of one or more rooms. Each circuit is protected from an overload of electricity by the circuit breaker in the service panel. If too many appliances on a circuit are used at one time, the circuit breaker opens the circuit. This prevents the wire in the circuits from overheating and starting a fire.

Fig. 12-27. Electrical roughing-in includes installing junction boxes and wires.

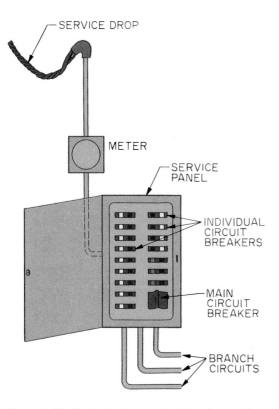

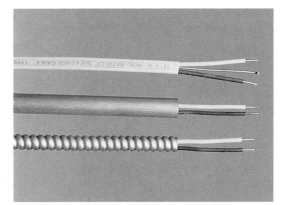

Fig. 12-29. Types of wiring from top to bottom: nonmetallic sheathed cable; conduit with individual wires; armored cable.

Fig. 12-28. Typical electrical wiring for residential service.

Plumbing Systems

Plumbing systems provide buildings with water and drainage. The plumbing system in a frame structure is installed within the wall, floor, and ceiling frames. In structures that are built on concrete slabs, some piping must be installed before the concrete can be poured. Fig. 12-30. Plumbing pipes are usually made of copper, galvanized iron, or plastic.

Fig. 12-30. In buildings that have a slab foundation, some plumbing must be done before the concrete can be poured.

Types of Wiring

There are three basic types of electrical wiring. Fig. 12-29. The type of wire used depends on local building codes, on the proposed use of the wire, and on the type of building being constructed. **Conduit** is a pipe through which individual wires can be pulled. It is used in concrete walls and floors to allow easy access to the wires for maintenance and repair. **Nonmetallic sheathed cable** is made of several wires wrapped together inside a plastic coating of insulation. It is very flexible and easy to install. **Armored cable** is made of several wires inside a flexible metal casing. It is used in dry environments where sturdy, yet flexible, wiring is needed.

Supply plumbing is the part of the plumbing system that supplies fresh water for drinking and washing. A service tap provides a supply of water from the city water main. A building supply line runs from the service tap to the water meter. Beyond the water meter, branch lines run to each location in the building where water is needed. Fig. 12-31. Because supply plumbing operates under pressure, the whole system must be leakproof. Valves are used to control the flow of the water.

Drainage plumbing is the part of the plumbing system that carries wastewater away from the structure. This part of the system is made up of drain and vent pipes. Fig. 12-32. Sometimes this part of the system is known as *DWV* (drainage, waste, and vent) plumbing. Copper, cast iron, and plastic pipes are used for drainage plumbing.

Did You Know?

Plumbing is not a recent invention. It was used in 2000 B.C. at the royal palace at Minos, in ancient Crete. Crete is an island in the Mediterranean. The royal palace had a system for bringing water into the palace. There also was a system for removing waste water. There also was a water-flushing toilet. Though it was simple, this plumbing system was effective. In its basic design, it was similar to modern plumbing systems.

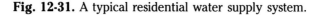

Fig. 12-31. A typical residential water supply system.

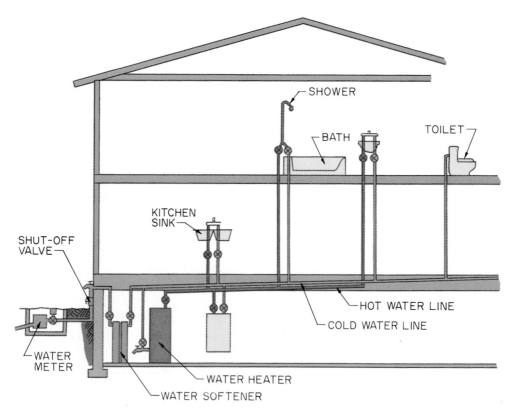

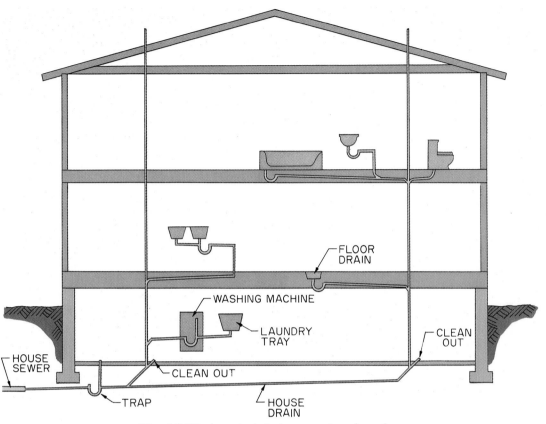

Fig. 12-32. A typical drainage system for a house.

Other Piping Systems

Pipes other than those used in supply and drainage plumbing are simply called *piping*. These pipes are used to carry natural gas, compressed air, steam, and water for fire protection. Most piping systems are installed by specialists called *pipefitters*. Pipes for these systems are available in a variety of materials and sizes. The type of pipe that is used depends on the purpose of the system.

HVAC Systems

The HVAC system controls the environment inside a building. This system consists of an air-handling unit, an air conditioner and furnace, and ducts. Installing this system is a very complex job. All of the parts have to work together properly. The installer needs to be familiar with the electrical, piping, and mechanical parts of the HVAC system.

The *air-handling unit (AHU)* is made up of a fan and a motor enclosed in a large, sheet-metal box. The AHU is the part of a HVAC system that moves the air through the building. As the air moves through the building, the *air conditioner and furnace* condition the air by changing its temperature. *Ducts* are the large pipes or tubes that carry the air to and from the air conditioner and furnace. *Supply ducts* carry air from the air conditioner and furnace to the rooms of the building. *Return-air ducts* carry the room air back to the air conditioner and furnace to be conditioned again. *Dampers* are valves that control the flow of air. *Thermostats* are automatic switches that measure temperature. They turn the system on or off to maintain a preset temperature range. Fig. 12-33.

For Discussion

Most stores and office buildings are now air conditioned. What effect has the use of air conditioning in buildings had on their general design?

Fig. 12-33. The thermostat measures the surrounding air temperature. It then switches the furnace or air conditioner on and off to maintain a preset temperature range.

FINISH WORK

After the utilities have been installed, the building is ready to be finished. Finish work completes the structure according to the building plans. It includes covering walls and ceilings, applying paint and wall covering, installing trim and hardware, paving and landscaping, and general cleanup. These final touches make the structure attractive and give the building a distinctive look.

Covering Interior Walls and Ceilings

Several materials may be used to cover interior walls and ceilings. The most common wall covering is drywall. Others include wood paneling and many different kinds of tile. Ceilings are usually covered with ceiling tiles or drywall.

Drywall is a general term used for plasterboard or wallboard. If both the ceiling and the walls are to be covered with drywall, first the ceiling is covered and then the walls. Drywall is fastened directly to ceiling joists and wall *studs*, or framing members. Next, the joints and nail holes are filled with drywall joint compound. The surface is then painted or covered with wallpaper.

Manufactured *wood paneling* often is used for interior walls. The panels are attached to the walls with nails or with an adhesive. Panels can be applied over drywall or nailed to furring strips attached to the studs. Drywall provides more uniform support for the panels than do furring strips.

Tiles are another common interior finish. Most wall tiles are made of ceramic. However, tiles can also be made of glass, steel, or plastic. Because they are easy to clean and maintain, tiles are used in restaurants, bathrooms, kitchens, and other places where cleanliness is important. Fig. 12-34. Wall tiles are applied

Fig. 12-34. Ceramic tile is used for walls that need to be cleaned often, such as walls in fast-food restaurants.

with adhesive to a special drywall backing. The tile joints are filled with *grout*, or cement, after the tile is in place.

Ceiling tiles are made of fiberboard or fiberglass. Some ceiling tiles are glued or stapled in place. Others are held in a suspended ceiling system. In a suspended ceiling system, metal strips are hung from the ceiling. The tiles are held in place by the metal strips. Fig. 12-35.

Painting and Wall Covering

Paint and wall covering are decorative finishes that are applied to interior walls. Both protect wall surfaces so that they last longer. They also make the walls easier to clean and maintain.

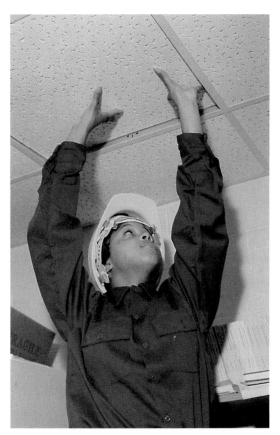

Fig. 12-35. A suspended ceiling system consists of tiles held by metal strips that are suspended from the ceiling.

Paint is a surface treatment that adds color to interior and exterior walls. It can be applied by brushing, rolling, or spraying. The kind of equipment used is determined by the size of the surface to be covered. Large surfaces are either rolled or sprayed. Fig. 12-36.

Wallpaper is an alternative to painting interior walls. Actually, most wallpaper is no longer made of paper. It is made of a thin layer of decorated vinyl. Wallpaper comes in many attractive colors and patterns. It is purchased by the roll and is cut to size by the installer. The pieces are applied to the wall with adhesive.

Did You Know?

In Europe wallpaper was developed in the late 1400s. The earliest European wallpapers were stencilled or painted by hand. The first wallpapers were meant to be decorative. They also were meant to resemble cloth wall hangings and wood paneling. It was only later that other wallpaper designs were introduced.

Installing Floor Covering

A floor covering is installed on the floors of most buildings. Floor coverings add to the beauty of the building. In addition, some types of floor covering provide cushioning against a hard concrete or wood floor. Floor tiles, sheet goods, and carpet are among the most common floor coverings. Fig. 12-37.

- *Floor tiles* are individual pieces of carpet, vinyl, or stone, usually 12 inches square. Adhesive is used to apply them to the floor.
- *Sheet goods* are large vinyl rolls, or sheets, of floor covering, usually 12 feet wide. The sheets are rolled out and cut to fit the room. Sheet goods usually are applied with adhesive.
- *Carpet* is a popular floor covering made of nylon, polyester, wool, or other fibers. Carpet comes in rolls that are 12 feet wide. The carpet is installed over a plastic or rubber-foam pad that acts as a cushion.

Installing Trim and Hardware

The last step in finishing the interior of a structure is installing the trim and hardware. Trim is installed around windows and doors. Special

Fig. 12-36. Paint can be rolled quickly onto large surfaces, such as the walls of this room.

trim called **molding** is used to cover joints where floors, walls, and ceilings meet. Fig. 12-38. This phase of building also includes installing cabinets, shelving, and other accessories. The hardware installed includes towel bars, soap holders, doorknobs, and shelf brackets.

At this point the second stage of the utility work must be done. Lights, fans, and other electrical fixtures are installed. Sinks, tubs, and faucets also are installed at this time. HVAC vents are put in place, and telephones are installed.

Finishing the Outside

After the structure has been completed, the outside work must be done to finish the site. There are four major kinds of outside finishing tasks: completing structural details, paving, landscaping, and cleaning up. Structural details

Fig. 12-38. Molding and trim are used to cover joints at the floor and ceiling and around doors and windows.

Fig. 12-37. Floor tiles, sheet goods, and carpet are three common floor coverings.

Fig. 12-39. The landscaping plan shows workers what to plant and where to plant it.

include tasks that must be done to finish the structure. The windows must be *caulked*, or sealed with a special compound. The exterior walls of most structures must be painted. (Stone and brick structures usually are exceptions.) Trim must be added, and porches and decks must be finished.

All of the paving is done at this time. Areas to be used for driveways, parking lots, and walkways need to be paved with asphalt or concrete.

First the earth is leveled and compacted. Then the pavement is laid in place and finished. Final paving work includes painting stripes for parking spaces.

Landscaping is done according to a landscaping plan, which shows how the finished site should look. It shows what is to be planted and where. Fig. 12-39. Trees, bushes, grass, and ground cover are planted in topsoil treated with lime and fertilizer to ensure better plant growth. Lawns are either seeded or sodded. Flower beds are planted to add beauty and color to the landscaping. Fig. 12-40.

The final step is to give the site a general cleaning. Any leftover debris is hauled away. The contractor makes a general tour of inspection and removes any tools and equipment that may still be at the site.

Fig. 12-40. Flower beds are an attractive highlight in any landscape plan.

For Discussion

Landscaping is an important step in finishing a construction project. In some cities, awards are given for beautiful landscaping. Is the land around your school landscaped? Can you suggest any improvements in the landscaping?

Construction Facts

STOP LANDSLIDES—USE SAWDUST!

Portions of Highway 55 near Boise, Idaho, are on a steep slope on the side of a mountain range. Landslides in this area are frequent and severe. They have been a recurring problem since 1946. Before this problem was solved, the roadbed in one 3½-mile (5.6-km) section kept slipping down the mountainside in eight different places.

In this particularly bad slide zone, engineers have used an unusual material to stop the slides: sawdust. Sawdust fill was used to stabilize the soil under the roadbed. Replacing the heavy existing soil with lightweight sawdust relieved some of the pressure on the water-saturated, unstable soil on the downhill side of the roadbed.

More than 33,000 cubic yards (25,000 cubic metres) of sawdust were dumped on the site. The sawdust was spread, mixed, and compacted. Then the regular rock roadbed, a seal coat of asphalt, and 1¼ inches (32 mm) of asphaltic concrete pavement were laid.

The decision to use sawdust was based on its success in other places. Also, the sawdust was available from a lumber mill 2½ miles (4 km) away. The lightweight material has done a good job of stabilizing the road. At the same time, the sawdust has been put to good use.

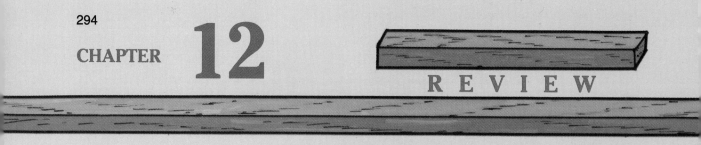

Chapter Summary

Construction processes are the processes used to build a construction project. The land on which the project will be constructed is called the site. The site must be surveyed and cleared. Laying out the site is the process of identifying the location of the proposed structure on the property. The excavation provides a place for the structure's foundation. The site must be leveled and graded. The soil must be stabilized to support the structure.

The foundation is that part of the structure that is below the first floor. The superstructure is the rest of the building. The footing is the part of the structure that distributes the structure's weight. The wall that is built directly on the footing is called the foundation wall. Forms are the molds that contain the concrete until it hardens. A slump test is made to test the workability of concrete. Screeding removes excess concrete and levels the top of the concrete. Floating is the process of moving coarse aggregate down into the concrete. The final smoothing is called troweling.

The three main structural parts of a building are the floors, walls, and roof structure. In bearing-wall structures, the walls support the weight of the building. In frame structures, the frame supports the weight of the building. The utilities are installed after the frame of the building has been enclosed. The service drop is wiring that connects a building to the electric company's wires.

There are three basic types of electrical wiring: conduit, nonmetallic sheathed cable, and armored cable. Plumbing systems supply buildings with water and drainage. The HVAC system controls the environment inside a building. Finish work completes the structure. Several materials may be used to cover interior walls and ceilings. The last step in finishing the structure is installing the trim and hardware.

Test Your Knowledge

1. What three steps are involved in preparing a site?
2. What are the markers called that identify a building's boundaries even after excavation has begun?
3. What name is given to a long, narrow excavation that is meant to hold pipelines?
4. Name three general methods of soil stabilization.
5. What are the two basic parts of a structure?
6. What three processes are used to smooth the surface of concrete?
7. In a bearing-wall structure, what supports the weight of the building?
8. What are the three main structural parts of a building?
9. What is a roof truss?
10. Name two types of roofing that can be used on flat roofs.

11. What are the three main types of utility systems?
12. What are the three basic types of electrical wiring?
13. What is the name of the utility system that controls the temperature inside a building?
14. Name three common types of floor covering.
15. What four things must be done to finish the area outside of a structure?

Activities

1. Find and observe a new building under construction. Report to your class on the type of structure it is.
2. Check the utilities in your house or apartment. Find out what utilities are available, where the meters are located, and who supplies each utility.
3. Choose a construction site in your area. Pretend that you are a contractor in charge of that building site. Plan what type of structure you will build. Put your plans on paper, showing what types of walls, flooring, roofing, and so on, you will use. Be ready to explain your decisions.

CHAPTER **13**

THE COMPLETED PROJECT

Terms to Know

certificate of
    occupancy
claim
final inspection
final payment
lien
maintenance

notice of completion
punch list
release of claims
release of liens
repair
warranty

Objectives

When you have finished reading this chapter, you should be able to do the following:

- Describe the final inspection process.
- Explain what a punch list is, who makes it, and what is done with it.
- List the steps to transfer ownership of a construction project.
- Explain how releases protect the owner.
- Explain the difference between maintenance and repair and give one example of each.

After a project has been completed, it is turned over to the owner. The owner makes the final payment to the contractor and takes responsibility for the structure. However, before the contract is considered officially fulfilled, several steps must be taken. This chapter will help you understand how a contract is concluded.

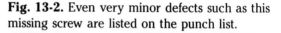

Touch up paint on living room wall.
Install threshold beneath garage access door.
Install switch plate in living room.
Touch up paint on bedroom wall.

Fig. 13-1. A punch list is made by the final inspection team to let the contractor know what corrections need to be made.

FINAL INSPECTION

Several inspections are made of a project while it is being constructed. The **final inspection** is made after a construction project has been completed. In this section, you will learn about the final inspection of a typical building. Every construction project must have a final inspection.

The building must be inspected to make sure that the job was done properly and according to the terms of the contract. The final inspection is made by several people working together as a team. The owner, the architect or engineer, and the contractor each has a representative on the team. The inspectors begin at the bottom of the building and work their way to the top. They look at everything. They inspect every room. They check to see that all mechanical and electrical equipment is working properly. They also make sure that everything has been installed safely. If, for example, electrical wiring was installed improperly, someone could be injured. The owner would be held responsible.

The Punch List

During the inspection, the inspectors list things they see that need to be corrected. This list is called a **punch list**. Fig. 13-1. Big problems, such as a crack in a ceiling, are on the list. However, the list also includes even the slightest defect, such as a missing screw or a mark on a wall. Fig. 13-2. When the inspection is complete, the punch list is given to the contractor. The contractor is then responsible for making the necessary corrections.

Fig. 13-2. Even very minor defects such as this missing screw are listed on the punch list.

HEALTH & SAFETY

Many of the inspection procedures on a new building are concerned with health and safety. Building inspectors are especially interested in making sure that the building will be safe for people to live and work in. They are concerned with the strength of materials. They also check that the correct construction techniques have been properly used. The laws regarding building inspection have become more strict in recent years. A hundred years ago, newly constructed buildings were not subject to the same careful inspection.

Corrections

Usually the inspectors give the contractor a *deadline*, or date by which the corrections must be completed. The contractor must evaluate the punch list and plan how each item will be corrected. The contractor must also determine who is at fault. If the contractor is at fault, company workers will make the corrections. If a subcontractor is at fault, that company must be notified that corrections are necessary.

After the corrections have been done, another inspection is made. This time, the inspectors check the corrected items closely. They also take another look around the structure. If the corrections are satisfactory and they do not find any new defects, the job is officially finished. However, if the corrections are not satisfactory or if new defects are found, the process is repeated. When everyone is finally satisfied that all of the contract requirements have been met, the owner formally accepts the building.

Certificate of Occupancy

If people are going to occupy the building, it must be approved by the city or county in which it was built. The building inspection department makes its own final inspection of the building. This is the same agency that issued the building permit that allowed construction to begin.

For a building to be approved for use, a **certificate of occupancy** must be issued by the building inspection department. This certificate shows that the building has passed the building inspector's check of the structure. It means that the building has been built according to federal, state, and local building codes. It means that the building is safe for people to use.

For Discussion

Assume that you are one of the inspectors making the final inspection on a building. What would you check in the building to make sure that the building was without defects? Describe the method you would use to make sure that you had carefully checked the entire building.

TRANSFERRING OWNERSHIP

After the inspections, punch list corrections, and final inspection have been made, the job is considered complete. The owner then files a formal **notice of completion**. This legal document lets everyone know that the job is finished.

Before the owner files the notice, however, he or she must receive certain documents from the contractor.

Releases

Before the property owner accepts ownership of the building from the contractor, the owner must get legal releases from the contractor. The two most common releases are the release of claims and the release of liens. These releases protect the owner from future claims for money and from liens on the building.

Release of Claims

A **claim** is a legal demand for money. A contractor may file a claim against the owner if the contractor runs into a problem that was not shown on the plans. For example, suppose that the contractor discovered poor soil on the site. The soil was not shown on the plans. It costs extra money to remove the poor soil and bring in better soil. The contractor did not expect to have to upgrade the soil. Thus, he or she did not include the cost of upgrading the soil in the contract price. Fig. 13-3.

To get a reasonable payment for the work, the contractor may file a claim for extra money from the owner. If the claim is not settled by the time the work is done, the claim is said to be outstanding. Both parties work toward an agreement so the claim can be settled. When an agreement cannot be reached, the claim must be settled in court.

Before the owner accepts the building, therefore, the contractor must provide the owner with a release of claims. A **release of claims** is a legal document in which the contractor gives up the right to file any claim against the owner. The contractor must check to be sure all claims have been satisfied before he or she signs the release of claims. If there are any unsettled claims, they are automatically cancelled when the contractor signs the release.

Fig. 13-3. In situations where unexpected problems are encountered, the contractor may have to file a claim.

Release of Liens

Any worker or supplier that has not been paid for materials or services can get a lien against the owner's property. A **lien** is similar to a claim, except that the owner's property becomes security for the amount due the worker. If the amount is not paid, the person who holds the lien can have the property sold. He or she can then take the amount due from the amount received from the sale. Thus, it is to the owner's advantage to make sure that the contractor has paid all the bills.

A **release of liens** is a formal, written statement that everyone has been properly paid by the contractor. The release provides assurance that no liens for unpaid bills can be put on the owner's property. When the contractor is satisfied that all the people involved in the project have been paid, then he or she can sign a release of liens. The owner then knows that the property is not in danger of being sold to pay a debt.

Warranties

A **warranty** is a guarantee or promise that a job has been done well or that the materials have no defects. When a project is finished, the contractor gives the owner certain warranties. The *contractor's warranty* is a promise that there are no defects in the work. The contractor promises to fix any flaws in workmanship within a certain time. This time is usually within one year of the issue date of the warranty. For example, if a door will not close properly, the contractor will fix it at no cost to the owner. The contractor wants to protect the reputation of the company. The warranty helps protect the construction company's good name.

The *supplier's warranty* is a guarantee that there are no defects in the materials supplied. Defects that might be found will be fixed by the supplier at no cost to the owner. Supplier's warranties cover items such as furnaces, air conditioners, and water heaters. Fig. 13-4.

When the construction is finished, the contractor gives the new owner a copy of the warranties. The contractor also makes sure that the owner receives the operating and service manuals for the equipment that has been installed.

Final Payment

As the owner receives the releases, warranties, and other items, he or she makes the **final payment**. This is the last step in transferring ownership. The owner pays the construction

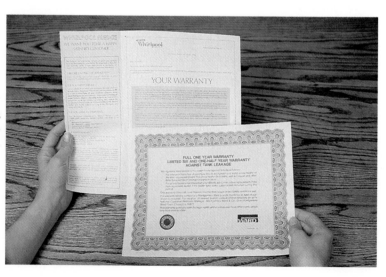

Fig. 13-4. This warranty states that the manufacturer will repair or replace any defective parts for a period of one year.

company all the money due according to the contract. Now the deal is closed and the owner can move in. The building has become the owner's responsibility.

For Discussion

From reading the text, you can see that warranties are important in building construction. In the past fifty years, buildings have become more complicated. Building contractors now install complex equipment in some buildings. One example of such equipment is a heating and ventilating system. Discuss the items that might need to be covered in such a product warranty.

MAINTENANCE AND REPAIR

The owner is responsible for the completed project, including all maintenance of the building and grounds. Most repairs are made initially by the suppliers and contractor under the construction warranties. However, after the term of the warranty is up, the owner is responsible for repairs.

Maintenance

Maintenance is taking care of an object or structure so that it continues to function the way it was intended. When you clean your room you are maintaining it. You put everything in its place. Then, the next time you look for your books, you will be able to find them. The maintenance your family does on your home includes keeping it clean and attractive inside and out. Fig. 13-5.

Some maintenance work needs to be done only occasionally, when the need arises. For example, when a light bulb burns out, you replace it. When the furnace filter gets dirty, you clean or replace it. Fig. 13-6.

Fig. 13-5. This homeowner is maintaining his house by checking the fit of screens and storm windows.

Fig. 13-6. Changing a furnace filter is an example of maintenance that needs to be done at certain intervals.

Other maintenance needs to be done on a regular basis. This type of maintenance is called *preventive maintenance* because it is done to prevent a problem from occurring. For example, a wood-frame house must be painted every three or four years to keep the wood from rotting and to maintain a neat appearance. The blower motor on a furnace or air conditioner must be oiled periodically to prevent the motor from failing.

Repair

Repair is the process of restoring an object or structure to its original appearance or working order. There are several reasons why repairs might need to be done. Many materials naturally *deteriorate* (wear out). They can also be damaged by extreme weather conditions and air pollution. Roofing materials, for example, wear out in time and must be replaced with new materials. Replacing broken or worn equipment is another kind of repair. For example, a dishwasher may wear out and need to be replaced.

Repairs may be minor and inexpensive. Repairs may also be major and require expensive, specialized help. For example, patching a hole in a wall is a minor repair. However, fixing an air conditioner or replacing a roof is considered a major repair. Fig. 13-7. In many cases, proper maintenance will help reduce the need for both minor and major repairs.

Maintenance is important in keeping a building in good shape. In the last twenty years, several materials have been introduced that reduce the need for building maintenance. Can you identify some of these new materials? What effects have the use of these new materials had on building maintenance practices?

Fig. 13-7. A special crew of workers is needed to repair this roof.

Construction Facts

THE ENERGY HOME

A house built in central Illinois is a showcase for many new developments in construction technology. Called the Energy Home, this 1,700 square foot (157 square metre), two-story house is a single family dwelling. It incorporates a number of unusual construction techniques. In time, however, some of the new ideas may become more common in the construction industry. For example, the foundation is made of plastic foam blocks. These plastic blocks were filled with concrete and reinforced with steel bars. The blocks were then covered with a waterproof coating.

The beams of the house are rectangular wooden trusses, rather than wooden beams. Such trusses were used because they could be easily made in the size and shape needed. It took only two days to enclose the Energy Home using the pre-cut trusses.

Heat is provided by a water heater/air handler system. This works in the same way as a traditional furnace. It is, however, more efficient. Air conditioning is provided by a gas-fired water chiller. This will reduce air conditioning costs. Natural gas is piped through semi-rigid stainless steel piping.

The windows were designed to reduce heating and air conditioning costs. The house also has been built along an east-west axis.

CHAPTER 13

REVIEW

Chapter Summary

Several inspections are made while a project is being constructed. After the project has been completed, the final inspection is made. During this inspection, a punch list is drawn up. This lists defects found during the inspection. These defects must be corrected by the contractor before a certain deadline. The owner accepts the building only when all of the contract requirements have been met. For a building to be approved for use, a certificate of occupancy must be issued. After the job is considered to be complete, the owner files a formal notice of completion. Before accepting ownership from the contractor, the owner must obtain legal releases. The two most common releases are the release of claims and the release of liens. A claim is a legal demand for money. In filing a release of claims, the contractor gives up any right to file any claim against the owner.

A lien is similar to a claim. In a lien, the owner's property is security for the amount owed the workers. A release of claims acknowledges that everyone has been properly paid by the contractor. A warranty is a guarantee that the job has been done well and is free of defects. A suppliers' warranty is a guarantee that there were no defects in the materials as supplied. The owner makes the final payment when he or she receives the release, warranties, and other needed items. Maintenance and repair are essential if the house is to be kept in good condition.

Test Your Knowledge

1. When is the final inspection of a construction project made?
2. Who makes the final inspection?
3. What is a punch list?
4. Who is responsible for seeing that all the needed corrections are made?
5. Who issues a certificate of occupancy?
6. Name two types of releases that an owner should get from a contractor before accepting ownership of the building.
7. What is a warranty?
8. Name two kinds of warranties.
9. What is the last step in transferring ownership from the contractor to the owner?
10. What is the difference between maintenance and repair?

REVIEW

 1. Visit a public building such as a library or courthouse. Find the certificate of occupancy posted on a wall somewhere in the building. You may need to ask to see it. Make a list of the conditions stated on it (maximum occupancy, etc.).

 2. Look around your house or apartment. Do the following:
   a. Make a list of things that need to be done as the need arises.
   b. Make a list of things that need preventive maintenance.

CHAPTER **14** STUDENT ENTERPRISE

Terms to Know

on contract
speculation
unit cost

Objectives

When you have finished reading this chapter, you should be able to do the following:

- Participate in the formation of a class company.
- Participate in estimating and planning a construction project.
- Participate in building a structure as a member of a class construction company.

You have studied construction and learned about tools, materials, and processes. You have learned how construction companies are organized and how they conduct business. You have also learned that much is involved in building any construction project. Now it is time for you to put into practice some of what you have learned. This chapter is designed to give you the opportunity to do some of the things you have read about. Most of the information in this chapter should be familiar.

Before you make any decisions about an enterprise project, you should read this chapter completely. Then you will know the facts needed to make wise decisions.

FORMING A CONSTRUCTION COMPANY

The development of your construction company will require the talents and abilities of every person in your class. To make sure that every student can take an active role in the company, you should form a corporation. You may wish to review the information about corporations in Chapter 7 before you begin to organize your company. Every student in your class should be a member of the board of directors. As such, each student will have full voting privileges and can participate in company decisions.

Company Organization

Successful companies are well organized. Each employee knows what he or she is supposed to do. Some people are managers. Others work with records and other paperwork. Still others do the physical work. Each worker is important. Each person must do his or her job well if the company is to be a success. Your board of directors will be responsible for organizing all of these activities.

You (the board of directors) should first elect the company managers. Fig. 14-1. You will need a project manager, a finance manager, a construction superintendent, a marketing director, and a personnel director. You will also need a manager for each part of the construction project. For example, there should be a manager

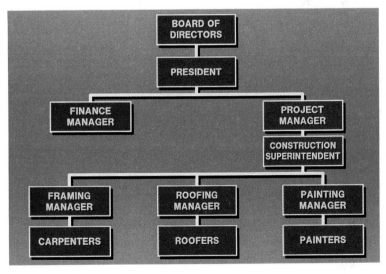

Fig. 14-1. A sample construction corporation organization chart.

for framing, one for roofing, and one for painting. The construction superintendent will coordinate the activities of the individual managers.

The marketing director will coordinate the advertising campaign and the sale of your project. He or she should be familiar with advertising techniques. The marketing director can also assume the responsibilities of finance manager. You can choose another student to handle your company's finances.

The personnel director will be responsible for hiring all the other students in the class. The personnel director should assign the jobs according to each student's desires and qualifications. The board of directors should approve each job assignment. Every student should be given a meaningful job. This will allow each student to see the dignity in honest labor. If there are too many jobs for the number of students available, some students may volunteer to do more than one job.

The personnel director must be sure that no one student is overburdened with responsibilities. This might prevent them from developing productive work habits. If there are too many students for the number of jobs available, the personnel director should consult the board of directors. The board of directors may vote on the best way to involve all the students in the company.

Company Organization and Leadership

The organization of a company provides opportunities to develop leadership skills. For example, the person in charge of a painting crew would need to direct those on the crew. Construction is a team effort. In any team effort, organization is important. Proper leadership helps organize a construction project.

One way to develop leadership skills is by joining a club. In your school, there probably are many clubs. Each club brings together people with an interest in one activity. For example, there are stamp clubs and speech clubs. There are also technology clubs. One such national organization is the Technology Student Association (TSA). It seeks to develop an understanding of technology.

The development of leadership skills will help you throughout life. Such skills will be valuable to you in any career you choose. The development of leadership skills will help you:
• To improve consumer understanding.
• To make good use of your leisure time.
• To recognize high standards of achievement.

Taking part in the activities of a club can help you develop leadership skills. Participation in club activities will help you learn:
• To prepare for effective citizenship.
• To be an officer in the organization.
• To prepare for effective participation in our democratic society.
• To conduct a meeting according to the rules of parliamentary procedure.

Taking part in group activities will also help you develop your social skills. These are the skills that help you get along with others. In any career you choose, it will be important for you to be able:
• To speak clearly.
• To write clearly and express your ideas precisely.
• To complete a job with or without direction.

The development of leadership skills is important. While in school, you should try to develop these skills as fully as possible. The organization of a construction company provides you with a good opportunity.

Project Financing

There are two ways your company can approach a construction project. You can build it on speculation or on contract. **Speculation** means that you build the project and then find someone to buy it. To build a project **on contract** means that you have a buyer before you start. You and the buyer, or owner, sign a contract. You build the project the way the owner wants it built.

If you can find an owner, or buyer, for the project your company intends to build, the owner will pay for the materials. However, if you build your project on speculation, your company will need money to buy materials for the project. Because your company is a corporation, you can sell shares of stock. The money raised from the sale of stock can be used to buy materials. Of course, the stockholders will expect a share of the profits after the project is sold. Another convenient way for your corporation to finance the project is to borrow the money. Perhaps your school will make the loan. Keep in mind, however, that the loan will have to be repaid after the project is sold.

For Discussion

Building a house on speculation is different from building a house on contract. The person who builds a house on contract already has a buyer. The person who builds a house on speculation does not have a buyer. Building a house on speculation is more risky than building a house on contract. The person who is considering building a house on speculation should consider several things before beginning the project. What are some of the things that he or she should consider?

THE CONSTRUCTION PROJECT

The construction project your company will build is a utility shed. Plans and specifications are provided here for a relatively low-cost, easy-to-build shed. The shed can be used to store firewood, garden equipment, bicycles, or almost anything. Fig. 14-2.

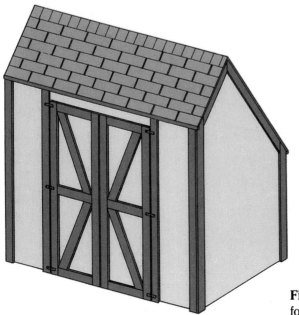

Fig. 14-2. This utility shed is an excellent project for your construction company to build.

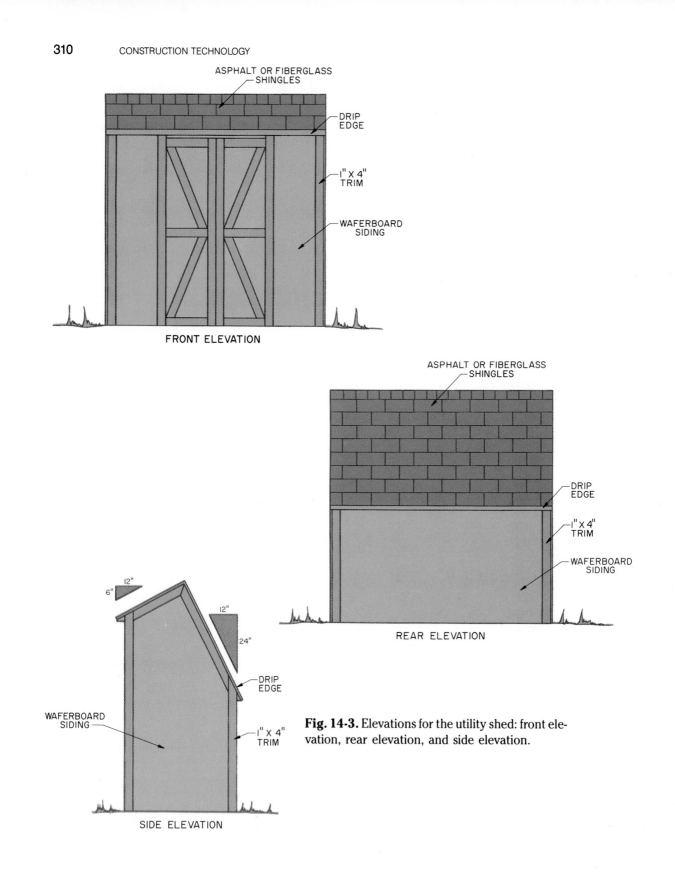

ASPHALT OR FIBERGLASS
SHINGLES

DRIP
EDGE

I" X 4"
TRIM

WAFERBOARD
SIDING

FRONT ELEVATION

ASPHALT OR FIBERGLASS
SHINGLES

DRIP
EDGE

I" X 4"
TRIM

WAFERBOARD
SIDING

REAR ELEVATION

12"

6"

12"

24"

DRIP
EDGE

WAFERBOARD
SIDING

I" X 4"
TRIM

SIDE ELEVATION

Fig. 14-3. Elevations for the utility shed: front elevation, rear elevation, and side elevation.

Plans and Specifications

Figure 14-3 shows the elevation views of the shed. General specifications are given in Fig. 14-4. Figures 14-5 through 14-13 show various details of the shed's construction. Study the plans and specifications carefully. Try to understand all of the parts and how they fit together.

Project Specifications

Size: 4′ × 8′ × approximately 8′ tall. 4′ door opening.

Framing: 2″ × 4″ pine or fir, 12d or 16d common nails.

Siding, flooring, and roof sheathing: 7/16″ waferboard, 6d common nails.

Trim: 1″ × 4″ pine or fir, 8d finishing nails.

Roofing: 15 lb. felt, aluminum drip edge, asphalt or fiberglass shingles, 1″ galvanized roofing nails.

Paint: exterior grade, flat, any color.

Fig. 14-4. Specifications for the utility shed.

Fig. 14-5. Shed floor framing plan.

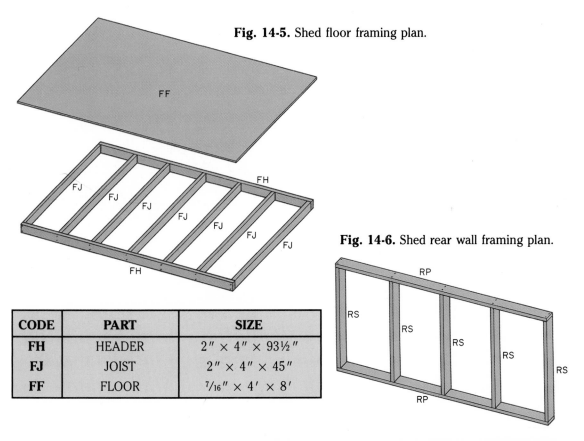

CODE	PART	SIZE
FH	HEADER	2″ × 4″ × 93½″
FJ	JOIST	2″ × 4″ × 45″
FF	FLOOR	⁷⁄₁₆″ × 4′ × 8′

Fig. 14-6. Shed rear wall framing plan.

CODE	PART	SIZE
RP	PLATE	2″ × 4″ × 93½″
RS	STUD	2″ × 4″ × 41″

CODE	PART	SIZE
SP	PLATE	$2'' \times 4'' \times 41''$
SS	STUD	$2'' \times 4'' \times 41''$

Fig. 14-7. Shed side wall framing plan.

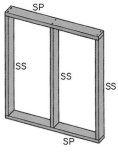

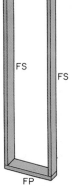

CODE	PART	SIZE
FP	PLATE	$2'' \times 4'' \times 21''$
FS	STUD	$2'' \times 4'' \times 74''$

Fig. 14-8. Shed front wall framing plan.

Fig. 14-10. Shed assembly detail.

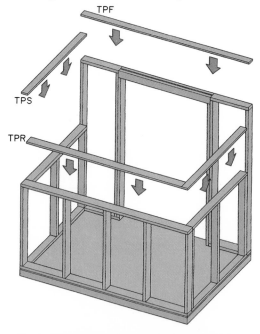

Fig. 14-9. Shed door frame detail.

CODE	PART	SIZE
DT	TRIMMER	$2'' \times 4'' \times 73\frac{1}{2}''$
DH	HEADER	$2'' \times 4'' \times 51\frac{1}{2}''$

CODE	PART	SIZE
TPF	TOP PLATE	$1'' \times 4'' \times 93\frac{1}{2}''$
TPR	TOP PLATE	$1'' \times 4'' \times 86\frac{1}{2}''$
TPS	TOP PLATE	$1'' \times 4'' \times 44\frac{1}{2}''$

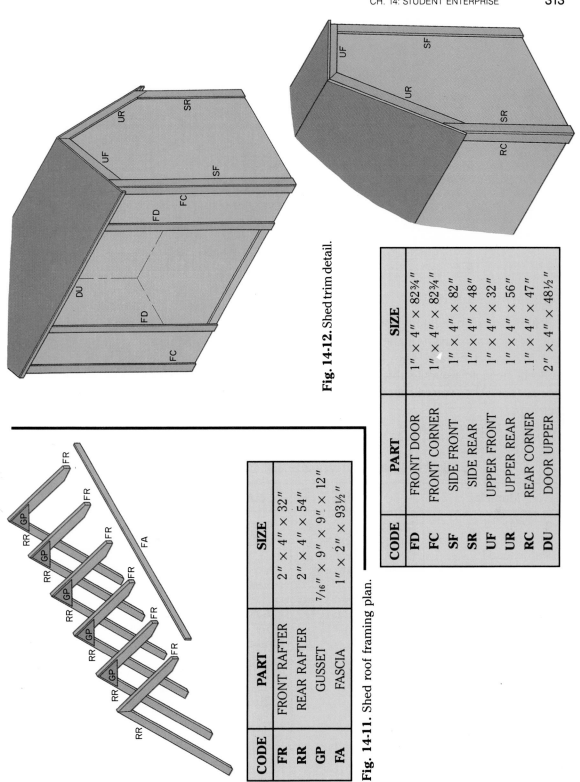

Fig. 14-12. Shed trim detail.

CODE	PART	SIZE
FD	FRONT DOOR	1″ × 4″ × 82¾″
FC	FRONT CORNER	1″ × 4″ × 82¾″
SF	SIDE FRONT	1″ × 4″ × 82″
SR	SIDE REAR	1″ × 4″ × 48″
UF	UPPER FRONT	1″ × 4″ × 32″
UR	UPPER REAR	1″ × 4″ × 56″
RC	REAR CORNER	1″ × 4″ × 47″
DU	DOOR UPPER	2″ × 4″ × 48½″

CODE	PART	SIZE
FR	FRONT RAFTER	2″ × 4″ × 32″
RR	REAR RAFTER	2″ × 4″ × 54″
GP	GUSSET	7/16″ × 9″ × 9″ × 12″
FA	FASCIA	1″ × 2″ × 93½″

Fig. 14-11. Shed roof framing plan.

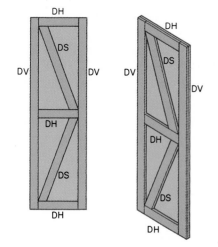

CODE	PART	SIZE
DD	DOOR	$^{7}/_{16}'' \times 24'' \times 79''$
DV	VERTICAL TRIM	$1'' \times 4'' \times 79''$
DH	HORIZ. TRIM	$1'' \times 4'' \times 17''$
DS	SLANTED TRIM	$1'' \times 4'' \times 40''$

Fig. 14-13. Shed door detail.

Estimating Materials

Before you begin building, you will need to order materials. To buy the materials, you must know how much of each is needed. Therefore you will need to estimate the quantities of the materials. Then you will need to determine their cost.

Your finance manager will be responsible for making a list similar to the one shown in Fig. 14-14. This list will help to determine the material quantities. Then study the plans and count how many of each type of material is needed. To make the estimate easier, round off all lengths to the next whole foot. Write your figures in the *quantity* column.

After the quantities have been determined, calculate the costs. You will need to know the unit cost for each material. **Unit cost** is the cost of the material per selling unit. When you find the cost of an item, write it in the *unit cost*

Fig. 14-14. Sample material takeoff and cost estimating form.

Material	Used for	Unit	Quantity	Unit cost	Extension
2″ × 4″	framing	LF			
1″ × 4″	trim, top plates	LF			
Waferboard	floor, siding, roof sheathing, doors	sheet			
Roofing Materials					
15-lb. felt	underlayment	SF			
drip edge	drip edge	LF			
shingles	roof	square			
Miscellaneous					
Nails					
16d common	framing	lb			
6d common	sheathing	lb			
8d finish	trim	lb			
1″ galvanized roofing	shingles	lb			
Hardware					
Hinges	doors	pair			
Hasp	doors	each			
Paint	siding	gal			
	trim	qt			

column. Construction materials are commonly sold in units such as linear foot (LF), square foot (SF), pound (lb), or square. In addition, some materials are sold individually. These materials are designated by the word *each* in the *unit* column. Once you know the unit cost, multiply it by the quantity. Write that amount in the *extension* column. Finally, add the dollar amounts in the *extension* column to get a total cost estimate for the project.

Planning and Scheduling the Work

To work efficiently, your company must have a work plan. This plan should include a list of the work to be done, who is to do the work, and when it is to be done. The whole project should be divided into small, manageable jobs. For example, the jobs may be floor framing, wall framing, roofing, and so on. Your project manager and construction superintendent should see that the work plan is made.

After you have identified the small jobs, you should decide on the order of work. Then you can begin to make job assignments. Obviously, not everyone can work on the same task at the same time. Therefore a schedule is needed. This is the job of your project manager. He or she should decide how much time each task should take, then make a large bar chart similar to the one in Fig. 14-15. Put the chart on a wall where everyone can see it.

The use of such a bar chart will encourage the construction team members to develop dependable work habits. It will also encourage them to be on time for their job assignments. Such a chart also will help build a spirit of teamwork.

Did You Know?

The first shelters constructed by humans probably were small, simple windbreaks. These may have been constructed from whatever materials were available. Reeds and branches from trees and shrubbery may have been used. These might have been fixed together by crudely interweaving them.

By stabilizing the windbreak and adding a roof to it, the windbreak became a crude lean-to. If the lean-to was enclosed, a simple shed was created. Each of these improvements resulted in a structure that offered more protection against the weather. As the shelter was completely enclosed, it also offered some slight protection against wild animals.

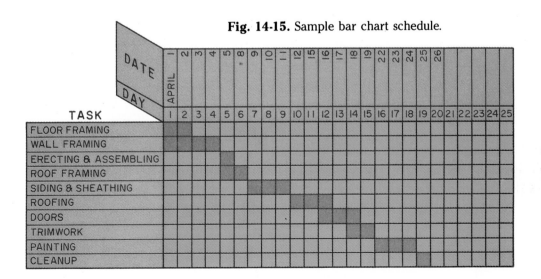

Fig. 14-15. Sample bar chart schedule.

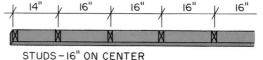

Fig. 14-16. Floor header layout.

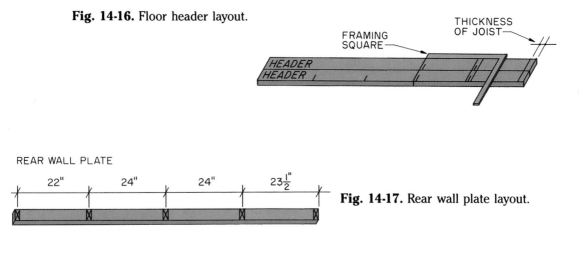

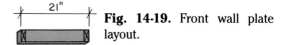

Fig. 14-17. Rear wall plate layout.

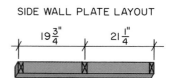

Fig. 14-18. Side wall plate layout.

Fig. 14-19. Front wall plate layout.

Building the Structure

Your project manager and construction superintendent are in charge of the project. You will need to follow their directions. Figures 14-16 through 14-39 show various construction details you will need to know. Also, you will need to refer back to Figs. 14-5 through 14-13.

Floor Framing

Cut the required lumber to the correct sizes. Mark the code letters on each piece as you cut them. Lay out the headers as shown in Fig. 14-16. Nail the headers to the joists using 16d common nails. Then nail the floor to the floor frame using 6d common nails spaced about 6 inches apart.

Wall Framing

Cut the required lumber to the correct sizes. Be sure also to cut the door trimmer studs and the door header parts at this time. Mark each piece with the correct code letters as you cut it. Lay out the top and bottom plates as shown in Figs. 14-17, 14-18, and 14-19. Nail the plates to the studs using 16d common nails.

Erecting and Assembling

Start by erecting the rear wall. Set it in place and align it flush with the edge of the floor. Fig. 14-20. Then nail it in place using 16d nails. Erect the side walls next using the same procedure. Then erect the front walls, again using the same procedure.

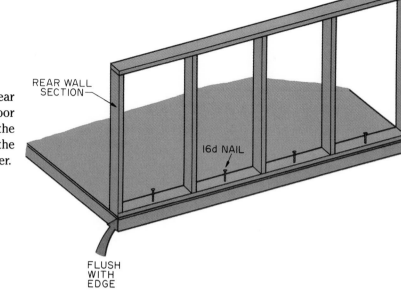

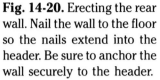

Fig. 14-20. Erecting the rear wall. Nail the wall to the floor so the nails extend into the header. Be sure to anchor the wall securely to the header.

After the walls have been fastened to the floor, they should be nailed together. Start at a rear corner. *Make sure the wall frames are square before you nail anything.* Nail the two walls together starting at the bottom and working up. Fig. 14-21. Repeat this procedure for all four corners.

Next, nail the door trimmer studs in place. Nail the door headers together with the waferboard spacer between them. Then nail the header in place on top of the trimmers. Fig. 14-22.

Finally, nail the double top plates in place. Be careful to align the walls as you nail the plates. Use 16d nails.

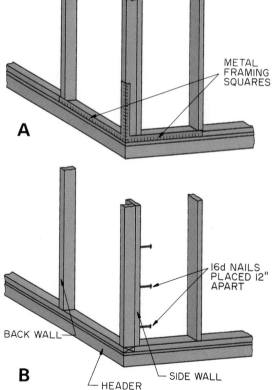

Fig. 14-21. Use a framing square to make sure the walls are square as shown in (a). Then nail the walls together as shown in (b).

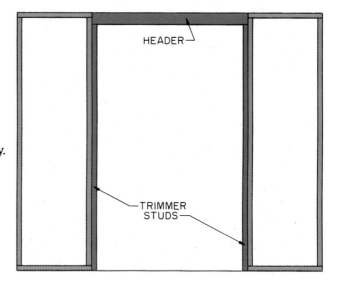

Fig. 14-22. Door frame assembly.

Roof Framing

Cut the required lumber to the correct sizes and angles. Figs. 14-23 and 14-24. Mark each piece with the correct code letters. Assemble the rafters and gusset plates using glue and 6d common nails. Use a framing square to align the parts while you assemble them. Fig. 14-25. Note that the end rafters need gusset plates on one side only.

Next, nail the rafters in place. The two end rafters should be flush with the end walls and the others should be 24 inches on center. Use 16d nails. Then nail the fascia board in place using 8d finishing nails. Fig. 14-26.

Siding and Sheathing

You will need three full sheets of waferboard. You will need one for the rear wall and one for each side wall. Use 6d nails spaced about 6 inches apart to attach all siding and sheathing. Begin by cutting the rear siding to size and nailing it in place.

Next, temporarily nail a sheet of waferboard to an end (side) wall. Use four nails and drive them only partly in. Mark the back side at the rafter line. Fig. 14-27. Then take the panel down and cut it to shape. Reinstall the siding and nail it in place permanently. Mark and cut the other end wall and nail it in place.

Measure both sides of the front wall and cut the siding to fit. Nail it in place on both sides. Note that no siding should be placed above the doorway.

Cut the roof sheathing to size. Nail the pieces in place. Fig. 14-28. Be sure that everything is aligned properly before you nail the parts permanently.

Trim

Cut the required lumber to the correct sizes. Note that the angles are the same ones used for the roof rafters. Refer to the elevation drawings in Fig. 14-3 for proper placement of the trim. Nail all the trim pieces in place using 8d finishing nails. Be careful not to damage the trim lumber. Refer again to Fig. 14-12.

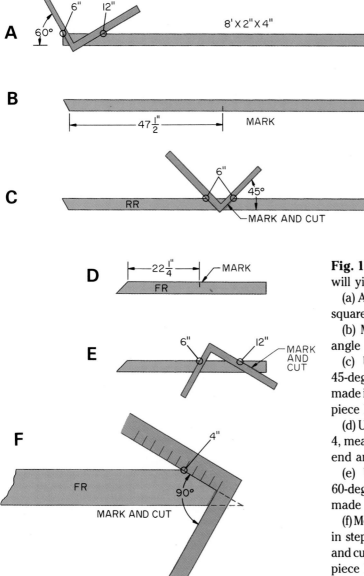

Fig. 14-23. Rafter layout. One 8-foot 2 × 4 will yield one RR and one FR.

(a) At one end of the 2 × 4, use a framing square to mark a 60-degree angle; then cut.

(b) Measure 47½″ from the base of the angle and mark that place on the wood.

(c) Use a framing square to mark a 45-degree angle, starting at the mark you made in step (*b*); then cut. This cut completes piece RR.

(d) Using the remaining portion of the 2 × 4, measure 22¼″ from the top of the angled end and mark that place on the wood.

(e) Use a framing square to mark a 60-degree angle starting at the mark you made in step (*d*); then cut.

(f) Measure down 4″ along the cut you made in step (*e*). Using the framing square, mark and cut a 90-degree angle. This cut completes piece FR.

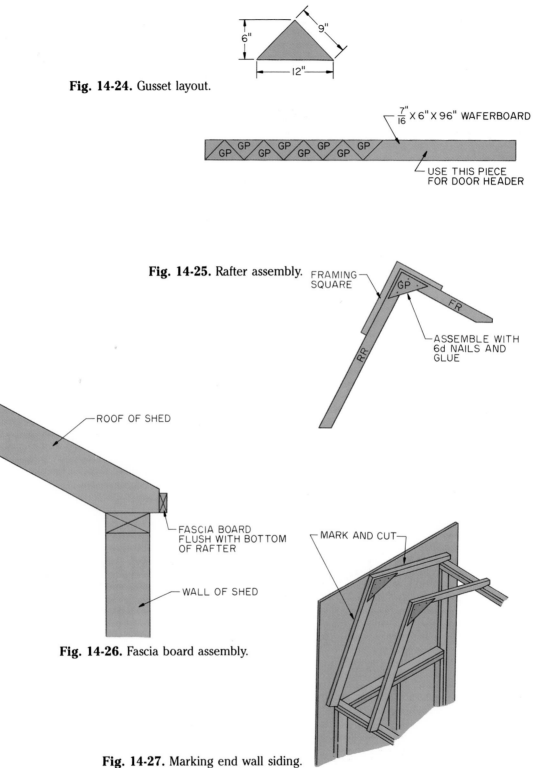

Fig. 14-24. Gusset layout.

$\frac{7}{16}$" X 6" X 96" WAFERBOARD

USE THIS PIECE FOR DOOR HEADER

Fig. 14-25. Rafter assembly.

FRAMING SQUARE

ASSEMBLE WITH 6d NAILS AND GLUE

ROOF OF SHED

FASCIA BOARD FLUSH WITH BOTTOM OF RAFTER

WALL OF SHED

Fig. 14-26. Fascia board assembly.

MARK AND CUT

Fig. 14-27. Marking end wall siding.

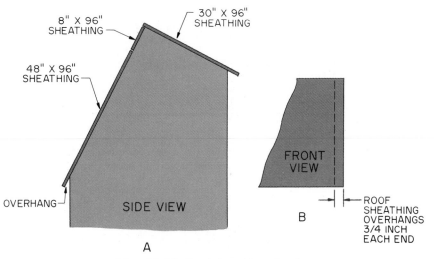

Fig. 14-28. Roof sheathing details.

Did You Know?

Most roofs today are covered with shingles. However, a variety of other roofing materials are available. These include clay roofing tiles, copper, and wood shingles. Throughout history, humans have chosen roofing materials from what was available to them. You may have seen pictures of cottages with thatched roofs. The thatched roof was used in rural areas of England, Scotland, and Ireland.

The thatched roof was used because it could be cheaply made from a material that was readily at hand—long grasses. These might be flax or rye. Today, the thatched roof is not common. It is, however, still found in parts of Ireland and Scotland. The thatch is applied as follows. The straw is laid in bundles on a sloped roof. The straw is then held down with ropes.

Roofing

Measure and cut the drip edge to length for the front and rear eaves. Nail the drip edge to the roof using roofing nails. Then staple the 15-pound felt in place. Fig. 14-29. Start at the bottom and work toward the top. Overlap each layer by about 2 inches. Overlap the pieces at the top.

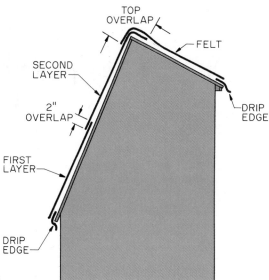

Fig. 14-29. Drip edge and felt detail.

Measure, cut, and bend the drip edge on the rake, or side, ends of the roof. This drip edge is nailed down *over* the felt. Fig. 14-30.

Next, install the shingles. For the starter course, turn the shingles so the tabs are toward the top. Start with a two and one-half tab shingle by cutting off half a tab. Nail this course as shown in Fig. 14-31. The purpose of the starter course is to strengthen the edge of the roof. Then place and nail the first course directly on top of the starter

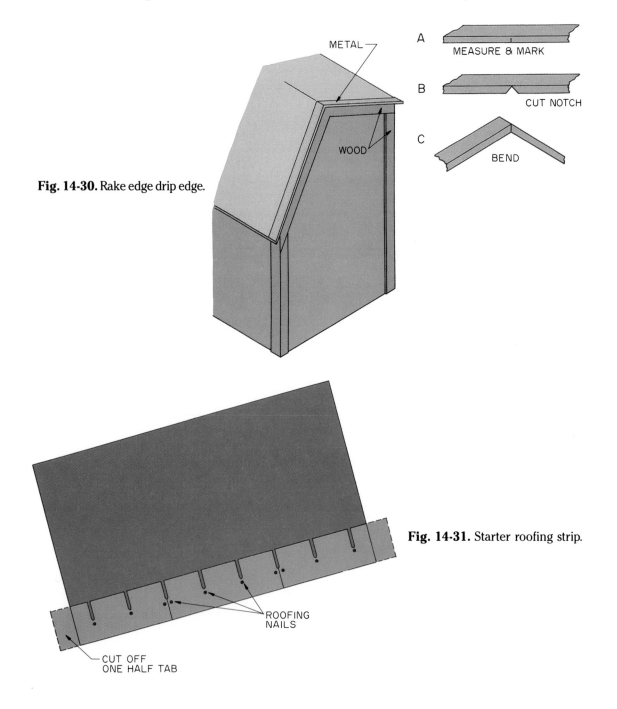

Fig. 14-30. Rake edge drip edge.

Fig. 14-31. Starter roofing strip.

course as shown in Fig. 14-32. Be sure that the tabs are facing toward the bottom. Place and nail the succeeding courses as shown in Fig. 14-33.

For the ridge, cut the shingles as shown in Fig. 14-34. Starting at one end of the ridge, nail the first shingle as shown in Fig. 14-35. Place and nail the other ridge shingles as shown in Fig. 14-36. The last shingle will need to be cut short and nailed carefully in place. Fig. 14-37.

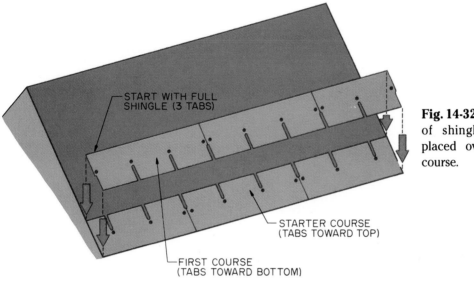

Fig. 14-32. The first course of shingles should be placed over the starter course.

Fig. 14-33. Succeeding courses of shingles. Note that the measurement increases by 5 inches for each course and that each starting shingle is ½-tab shorter than the preceding one.

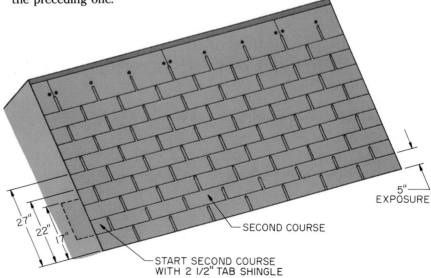

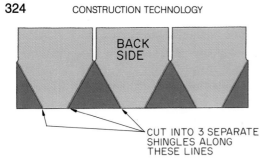

Fig. 14-34. Cutting ridge shingles.

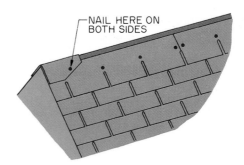
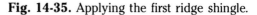

Fig. 14-35. Applying the first ridge shingle.

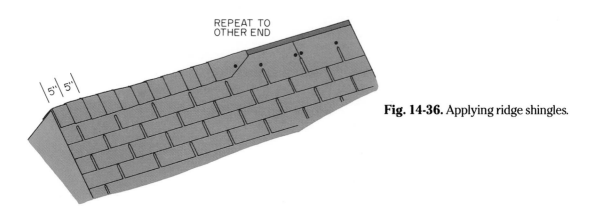

Fig. 14-36. Applying ridge shingles.

Did You Know?

In ancient times, doors were flexible. They were, for example, made of hanging hides or cloths. The use of rigid materials—such as wood—for doors was a later development. Stone doors were used by both the Greeks and the Romans. These doors opened on pivots placed at the top and bottom. However, the most common rigid door was the wooden door. Generally, these were quite simple. In design, many of them were quite similar to modern wooden doors.

In the middle ages, the most common type of wooden door was made of vertical boards. These boards were braced on the back with diagonal or horizontal braces.

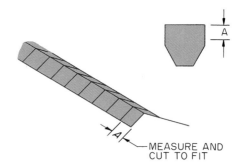

Fig. 14-37. Applying the last ridge shingle.

Doors

The doors require one full sheet of waferboard. Cut the sheet in half lengthwise. Then measure and cut each piece to the correct length. Next cut the door trim pieces. Attach the trim pieces to the waferboard using glue and 8d finishing nails. Fig. 14-38. *Clinch*, or bend over, the nails on the back side.

Fasten the hinges to the doors. Then lift one door into place. Mark the hinge hole locations on the front door trim (FD). Be sure to leave a space of $1/8$ inch between the door and the side of the building. Leave a $1/4$-inch space between the door and the top of the door frame. Fig. 14-39. Do the same for the other door. Then install the hasp and loop.

Painting

Put a first coat of paint on all bare exterior wood surfaces. Allow the paint to dry thoroughly. Then apply a second coat to the siding and doors. Let that coat dry. Finally, paint the trim pieces with a contrasting color. Be careful not to get trim paint on the siding.

Did You Know?

The first paints were used primarily for decoration. You may have heard of the cave paintings at Lascaux, France. These paintings on rock were made about 17,000 years ago. They are decorations on the cave walls. They show some of the animals familiar to early humans. In the 1500s in Europe, paint began to be used to protect and preserve objects.

Paint then was made by hand. It was expensive. It was only in the 1800s that paint became widely available—and cheaper. In the twentieth century, manufacturing increased. Many of the items manufactured needed painting. This led to growth in the paint industry. Today, a wide variety of paints is available. There are paints for many uses.

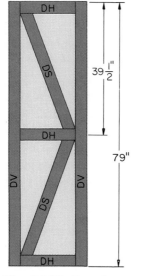

Fig. 14-38. Door layout.

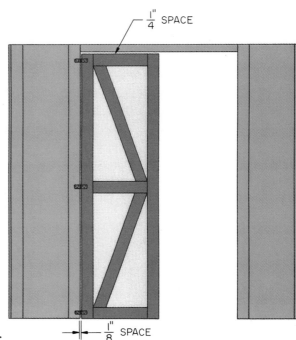

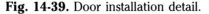

Fig. 14-39. Door installation detail.

For Discussion

As you can see, building any structure requires a number of different construction steps. Many of these basic steps have not changed in the last one hundred years. Only the tools needed may have changed. One other thing that may have changed is the time schedule. Many buildings today are built to meet a tight deadline. Discuss the effects of these changes on the need for careful organization of the steps in building construction.

DETERMINING A PROFIT OR LOSS

After the project has been completed, your company must sell the project if it has not already been sold. The project must also be delivered to the new owner. Then your company will need to account for the money it spent and received.

Marketing

If you built the shed on contract, it was sold before you started. If, however, you built the shed on speculation, you must find a buyer. The first task is to decide on a fair price for your project. Then you will need to let people know that you have a shed for sale. Possible ways to advertise are as follows:

- Post notices on school bulletin boards.
- Ask a local radio station to broadcast the availability of your product.
- Advertise in a local newspaper by having the paper publish a news story about your class.
- Another possibility is to try to sell your product by *word of mouth*.

Financial Accounting

The last step in your construction enterprise is to prepare a financial accounting. This will be your profit and loss statement. It will show in detail all the money the company received and spent. Figure 14-40 shows a sample profit and loss statement. If you sold shares of stock, any profit you make should be divided among the stockholders. If you borrowed the money for your project, you will need to pay back the amount you borrowed plus interest. Any amount that remains after the loan is repaid should be divided among the stockholders.

Fig. 14-40. Sample profit and loss statement.

For Discussion

The text suggests several methods for marketing your product. Analyze these various methods. Which of them, in your opinion, would be most effective in your community?

Construction Facts

A HOUSE IN THE WOODS

"Near the end of March, 1845, I borrowed an axe and went down to the woods by Walden Pond, nearest to where I intended to build my house, and began to cut down some tall arrowy white pines . . . for timber."

This begins Henry David Thoreau's account of how he built his own house. One of our great writers, Thoreau was an independent thinker who liked solitude. His goal was to live simply and economically.

Here is what he says about his work:

"I hewed the main timbers six inches square, most of the studs on two sides only, and the rafters and floor timbers on one side, leaving the rest of the bark on, so that they were much stronger than sawed ones."

"At length in the beginning of May . . . I set up the frame of my house . . . I began to occupy my house on the 4th of July.

"I have thus a tight shingled and plastered house, ten feet wide by fifteen long, with a garret and a closet, a large window on each side, one door, and a brick fireplace."

Thoreau was to live in the house year-round for two years and two months. When asked about the value of his experience for the young people of his time, Thoreau said: "I mean that they [students] should not *play* at life, or *study* it merely, while the community supports them at this expensive game, but earnestly *live* it from beginning to end."

CHAPTER **14**

R E V I E W

Chapter Summary

The development of your construction company will require the talents of every person in the class. In your company, you will need a project manager, a finance manager, a construction superintendent, a marketing director, and a personnel director. You will also need a manager for each major part of the project. You can build the project on speculation or on contract. In either case, you will need to make a careful estimate of the materials you will need. You also will need to develop a work plan.

The actual building of the shed is divided into several work areas. These are floor framing, wall framing, erecting and assembling, roof framing, siding and sheathing, work on trim, roofing, and doors, and painting. After completing the project, you will need to market it. You will then need to determine your profit or loss. You will do this by preparing a financial accounting.

Test Your Knowledge

1. Which corporation official is responsible for hiring employees?
2. What does it mean to build a project on speculation?
3. What is meant by unit cost?
4. Identify one way to develop leadership skills in school.
5. What is the purpose of a bar chart in scheduling work?

Activities

1. The organization of any project requires leadership. Using resources in your library, research the lives of two great leaders. Then, compare the lives and achievements of these leaders. Look especially for qualities that they shared. Using this comparison, identify those personal qualities that helped each of these individuals develop leadership.

2. After completing the student enterprise, identify those qualities essential for teamwork. Think about your role in the work team. Think about how your job related to the jobs of others on your team. Then write a one-page essay on the qualities that you think are essential for teamwork on a group project.

3. In your work on this student enterprise, you may have set some personal goals for yourself. For example, you may have set for yourself a goal of accomplishing so much work in a single class session. Proper management can help you reach personal goals on the job. List some ways in which proper management can assist you in reaching personal goals.

REVIEW

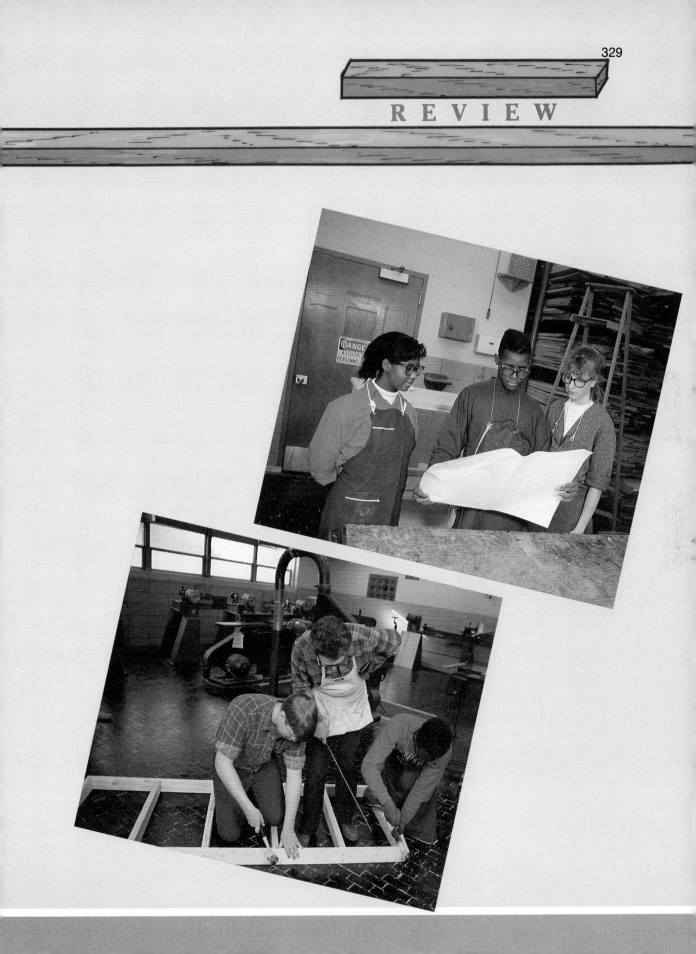

ACTIVITIES

Activity 1: Building and Testing a Model Bridge

Objective

After completing this activity, you will become familiar with the basic design features of today's basic bridge structures. The bridge models produced can also be used to test the strength of each support system. You also will become familiar with the basic principles of materials testing.

Materials Needed

For this activity, you will be working with a teammate. The materials list below assumes that there are ten teams of two students each.

- 120 $\frac{1}{8}$" × $\frac{1}{4}$" × 36" strips of balsa wood (12 per team)
- 10 $\frac{1}{16}$" × 3" × 36" strips of balsa wood (1 per group)
- 10 containers of glue (1 per group)
- 10 spools of thread (1 per group)
- 100 1" wire brads (10 per group)

Steps of Procedure

1. Cut all balsa wood strips into 12" lengths.
2. With your teammate, decide on the type of bridge you intend to construct. You may get your ideas from local bridge construction in your area. You might also research bridge design in your school library. Figure A shows the basic types of bridges.
3. After you have chosen a bridge, you should research that type of bridge. You should research its historical development. You also should attempt to find out the location of a similar bridge.
4. After you have identified the type of bridge you plan to build, you should make a working drawing of the bridge. With your teammate, you should construct a model of the bridge you have chosen. Each model should have an overall size of 12" × 3" × 6".
5. After you have completed the construction of your bridge, share your research information with the other teams.
6. To test the strength of your bridge, place two tables about one foot apart. Span the bridge between them.
7. Place a cardboard box beneath the bridge to protect the floor.
8. To test the strength of the bridge, place gym weights on the bridge—one at a time. Place the lightest weights first. Gradually add more weight. Record the amount of weight each bridge can hold before it collapses.
9. What type of bridge was able to hold the most weight? Why do you think some bridge designs are stronger than others.

ACTIVITIES

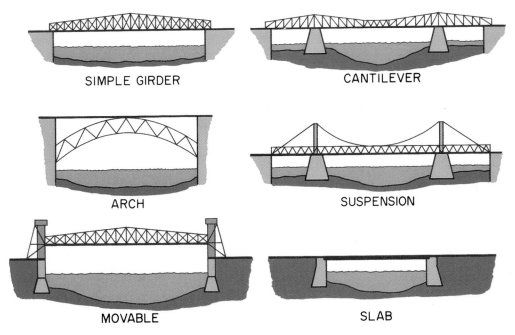

Fig. A. The main types of bridges.

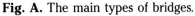

Activity 2: Building a Model Lock-and-Dam System

Objective

After completing this activity, you will be familiar with a basic lock-and-dam system. This is the type of system used in canal and river transportation, irrigation, and flood control. You also will have a basic knowledge of the plumbing skills used while working with plastic water pipe.

Materials Needed

- 3' × 1' × 1' watertight container. (This may be made of sheet metal.)
- 4 ⅜" NPT/female adapters
- 3 ⅜" compression Ts
- 3 ⅜" compression water valves
- 5' ⅜" plastic water line pipe
- 2 ⅜" 90° elbows
- 1 garden hose
- 2-1' × 1' × 1" pieces of cork board. (These pieces may also be the height and width of the water container.)
- 1-2" × 4" × 6" wood block. (This is needed for the barge.)

- 1 permanent colored marker
- 1 tube clear silicone caulking
- 1 small container of PVC glue
- 1 hacksaw

Steps of Procedure
Constructing the Lock and Dam

1. Divide the water tank into three equal sections across the bottom of the tank. Mark with the permanent marker. Fig. A.
2. Cut a 2″ strip from the 1′ × 1′ cork board. Place the strips over one of the dividing lines. Fig. 1.
3. Cut the 10″ × 12″ cork board (from Step 2) into two 10″ × 6″ pieces. Place these two halves on top of the 2″ strip to form Gate 1. Fig. A.

4. Cut the remaining 1′ × 1′ cork board in half. Place the two halves on the second dividing line to form Gate 2. Both Gate 1 and Gate 2 should fit snugly to prevent leakage. Fig. A.
5. Drill a ⅜″ hole in the bottom of each compartment.
6. Drill a ⅜″ hole in the end of the tank nearest to Gate 2. Drill the hole 2″ up from the bottom.
7. Place silicone on the threads of each adapter. Fasten each of the four adapters in the four ⅜″ holes just drilled. The silicone should act as a gasket to seal these fittings. Fig. B.
8. Cut a 1′ section of ⅜″ plastic water line. Attach it to the end adapter using PVC glue. This stem will help simulate the continuous flow of water moving downriver.
9. Using the compression valves, Ts, elbows, PVC glue, and ⅜″ water line, connect the water flow control system as shown in Fig. B.

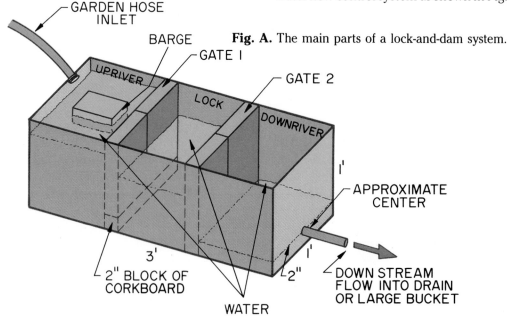

Fig. A. The main parts of a lock-and-dam system.

GARDEN HOSE INLET

BARGE

GATE 1

GATE 2

UPRIVER

LOCK

DOWNRIVER

1′

APPROXIMATE CENTER

3′

2″ BLOCK OF CORKBOARD

1′

2″

DOWN STREAM FLOW INTO DRAIN OR LARGE BUCKET

WATER

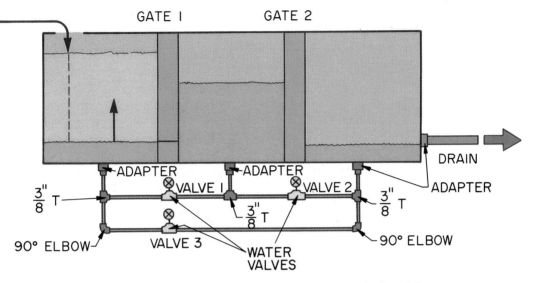

Fig. B. A side view of the water flow control system in a lock-and-dam system.

Steps of Procedure
Operating the Lock and Dam

1. Refer to Fig. B. Close all valves.
2. Run a garden hose to the "up-river" compartment. Turn on the water. Adjust it to a medium rate of flow.
3. Place the barge in the "up-river" compartment.
4. Open Valve 1 all the way. The lock will begin to fill.
5. As the first two compartments fill with water, you will need to open Valve 3. You will need to do this to keep a flood from occurring.
6. Once the first two compartments have balanced water levels, you may open Gate 1 and drive the barge into the lock.
7. Close Gate 1. Close Valves 1 and 3. Open Valve 2. The lock and down-river compartments should now be in balance.
8. Open Gate 2. Drive the barge downriver to allow the barge to pass out of the lock.

Activity 3: Building a Concrete-Block Wall

Objective

After completing this activity, you will have become familiar with the process of laying block.

You also will have learned how to mix mortar and build a wall that is straight and level.

Materials Needed (Refer to Fig. A on the following page.)

• 9 concrete or cinder blocks
• 1-10-lb. sack of cement

- 5' of string
- 1 gallon bucket of water
- 20 lbs. of sand
- 1-3' × 3' × ½" sheet of plywood
- 2 trowels
- 1-36" level
- 1 chalk line
- 1 strike-off tool

Steps of Procedure

1. Find a fairly level surface in your lab area. A smooth concrete floor is best.
2. Clean the edges and surfaces of all the concrete or cinder blocks.

3. Mix the cement on the 3' × 3' piece of plywood. This will serve as a mortar board. Using the trowel, mix five parts (scoops) of sand with three parts of cement. Mix the sand and cement on the mortar board. Work the mixture until it has an even gray color.
4. Add water to the mixture. Add only a small amount at first. Mix the water with the sand cement mixture. Continue to add water and mix the mortar until the mortar can be scraped to the middle of the mortar board and still hold its shape. The mortar should not be runny. It should hold its shape when scooped up with a trowel. When dropped on the mortar board, it should flatten slightly.

Fig. A. Masonry tools and materials. The items shown here will be needed to build the wall described in this activity.

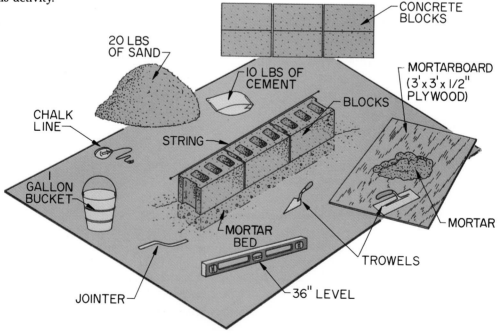

ACTIVITIES

5. Strike a chalk line about 4′ long on the floor where you plan to construct your wall.
6. Using your trowel, set a continuous bed, or layer, of cement along the edge of the chalk line. This bed should be approximately 2″ wide and 1″ high.
7. Set a second bed parallel to the first at a distance equal to the width of the block.
8. Lay three beds of mortar across the parallels to rest the first block on.
9. Set the first block in position. Tap it down into the mortar with the handle of the trowel. The mortar should be about 3/8″ thick. Strike away any excess mortar from the edge of the block. Mix this mortar back into the cement.
10. Lay three more beds of mortar for the second block. Set the second block and tap it into

position. Using a level, check that the tops of the blocks are level. Check that the mortar has a uniform thickness.
11. Check that the wall is straight. Do this by holding the string at the ends and corners of the tops of the two blocks. To align the blocks with the string, tap them in or out.
12. Continue this process with the third block. You will then have completed the bottom level of your wall.
13. Add the two remaining levels to the wall.
14. Ask your instructor to check your work before it completely dries.
15. Disassemble the wall and clean all of the blocks.
16. Rinse all tools in clean water. Do not wash your tools in a sink or it will become clogged.

Activity 4: Constructing a Model Shed

Objective

After completing this activity, you will know how to construct a simple model to represent the shape and size of a fully constructed shed.

Materials Needed

- 1-12″ × 18″ × 1/8″ sheet of foam board (available in local art supply stores)
- 1 sharp, pointed model knife
- 1-12″ metal straightedge to guide the cutting knife
- cutting board about 12″ × 12″
- 1-12″ scale
- drawing board and T-square

Steps of Procedure

1. The shed for which you will construct a model is described in full detail in Chapter 14. Figures A, B, and C provide you with the basic dimensions of the shed. The shed is 94″ long and 48″ wide. Figure A shows the construction of the shed walls.
2. The pitch of the roof is critical to the overall appearance of the shed. Place the foam board on the drawing board and measure 3″ (1/16″ equals 1″). Therefore, 3″ is composed of 48 1/16″ units. Use the triangle to extend a light pencil line vertically from the 3″ mark.
3. Next, measure up from the bottom a distance of 3″ to represent the shortest height of the shed (48″). On the left side measure up a distance of 5⅛″ to represent the tallest side of the shed (82″).

SECTION **IV**

ACTIVITIES

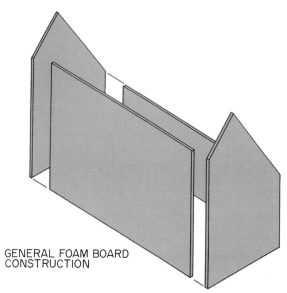

GENERAL FOAM BOARD
CONSTRUCTION

Fig. A. The construction of the shed walls.

4. Refer to Fig. B to calculate the pitch of the roof. The slope from the left has a 6″ rise for each 12″ of movement to the right. This has been calculated to have an angle of 36½°. The angle from the right is steeper, with a rise of 24″ for each 12″ of movement to the left. This has been calculated to have an angle of 62½°. You can construct the pitch for each side or use a protractor to create the angles. You will need to trace these two pieces on the foam board.

5. The front and rear of the shed are fairly easy pieces to make. The front of the shed has an overall measurement of 82″ × 94″. The rear of the shed measures 48″ × 94″. Fig. C. Converted to the scale used in constructing the sides, this means the front will mea-

sure 5⅛″ high × 5⅞″ long. The rear will measure 3″ high × 5⅞″ long.

6. After marking these four pieces on the foam board, cut them out. Put the foam board on your cutting board. Place a metal straight-edge along each line. Carefully insert the cutting knife into the foam board. Keep the knife straight up and down while slowly pulling it along the pencil line. Take care not to cut across the line into another section of the shed. **NOTE:** Hold the straightedge firmly and keep your fingers back from the blade.

7. When you have cut out all four pieces, hold them together. Check that they have the general shape of the shed. Measure the distance across the front of the shed. Let the ends of the shed be the outside pieces. Fit the front between these two ends.

8. You will discover the shed is actually 6⅛″ in length, rather than 5⅞″. The reason for this difference is the thickness of the foam board. To obtain an accurate representation of the shed, you will need to cut ¼″ from the length of the front and rear sections.

9. Begin construction by applying a small amount of hot glue along one edge of the front shed section. Place the tallest portion of an end section against the glue. Hold it in place until the glue hardens. Be certain to hold the section straight up and down.

10. Next, place glue along the other end of the front section. Attach the other end in the same manner. Since the glue hardens very quickly, do not put the glue on both ends at the same time.

11. Repeat the process for the rear section.

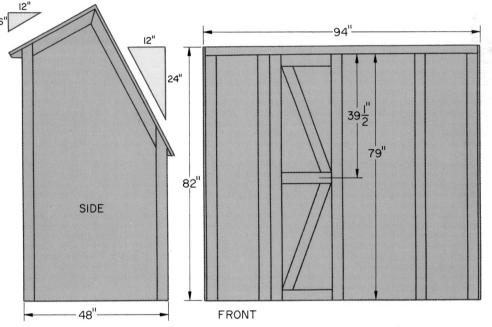

SIDE

FRONT

Fig. B. Panel dimensions and roof pitch.

12. Measure the distance of the longest slope of the shed roof. The distance should be 3⅛″ (50″) in length from the peak to the rear wall of the shed. This roof will also have an overhang of ⅛″ (2″) in the rear and on both sides. Cut a piece of foam board 3¼″ (52″) × 6⅛″ (98″).
13. Glue the rear roof section into place.
14. Measure the distance from the top end of the rear roof section to the front edge of the shed. Once again, allow ⅛″ (2″) of overhang along the front and on both ends. The piece of foam board should measure 2″ (32″) × 6⅛″ (98″). Glue this section in place so the front section of the roof overlaps the leading edge of the rear roof section.
15. Your model can be further refined by painting the trim on or by cutting the doors.

Fig. C. The rear panel.

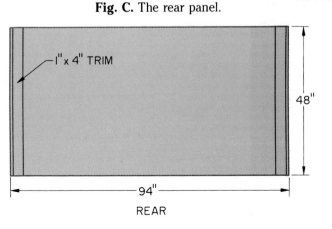

1″ x 4″ TRIM

48″

94″

REAR

SECTION

V

CAREERS

Chapter 15: Preparing for Construction Careers
Chapter 16: Careers in Construction

CHAPTER

15 PREPARING FOR CONSTRUCTION CAREERS

Terms to Know

ability
apprentice
aptitude
bachelor's degree
career
community college
*Dictionary of
    Occupational
    Titles (DOT)*

general education
job
*Occupational
    Outlook Handbook*
on-the-job training
technical institute
values

Objectives

**When you have finished reading this
chapter, you should be able to do the
following:**
- Know where to find information relating
  to construction careers.
- Identify careers related to construction.
- Know the different sources of training for
  construction-related careers.

One of the most important decisions you will ever have to make is the decision regarding the career you want to pursue. *Career* is not simply another word for *job*. A **job** is a paid position at a specific place or setting. A **career** is a sequence of related jobs that a person holds throughout his or her working life. Fig. 15-1. Your choice of career may affect every aspect of your future. Your relationships, your lifestyle, and even your health depend to a great extent on your choice of a career.

Fig. 15-1. A career is a sequence of related jobs. A carpenter, for example, might work his way up from laborer to construction superintendent over a period of many years.

The field of construction provides many different career opportunities for people with different interests. As you study construction technology, you may find an area of construction that interests you.

DETERMINING YOUR CAREER INTERESTS

Choosing a career and planning how you will prepare yourself for it can be confusing. There are so many exciting possibilities from which to choose! Do not be disappointed if you cannot decide on a career path in just a few weeks. Also, do not feel that you must make a career decision now. Rather, you should start exploring career options now. While you explore, you should be getting a basic education. A basic education is necessary for almost any career you choose. Fig. 15-2.

Fig. 15-2. Almost every career now requires a high-school diploma. Many careers even require a minimum of a college degree.

The years you spend in school can be some of the most exciting years of your life. Every week you learn new ideas and face new challenges. Many of these ideas and challenges are designed to provide you with the basic skills you will need after your school years are finished. You may not find all of the skills and information you receive interesting or of use to you right now. However, do not underestimate the value of this knowledge. As you grow older, you may be surprised at how well a good education can serve you.

Learning to Make Choices

The best way to go about making a career choice is to learn about all the possibilities. You will want to make a wise choice. Your choice of career, after all, will affect you for many years to come. Wise choices are choices that are made on the basis of *knowledge*.

At one time, most of the decisions in your life were made for you. The clothes you wore and the food you ate were chosen by your parents. You were not given any choices. Then, as you grew older, your parents began to ask your opinion about clothes, food, and other necessities. You began to choose food and clothing you knew that you liked. You based your choices on past experience.

Later in life, you started to make decisions without having first experienced the choices. When selecting courses in school, for example, you probably selected some that you were not familiar with. In these cases, you could not rely on experience to help you make your decisions. Instead, your decisions were probably based on information you received from others. You may also have found information in the course guide. In other words, you based your decisions on research.

Choosing a career also requires research. Few people are able to make career decisions based on experience in the career. In most cases, people must make career decisions long before they actually work in the career. Because of this, it is most important that you gather information now so that when the time comes to make a career choice, you will be prepared. Fig. 15-3.

Fig. 15-3. Many schools hold a Career Day once a year to help students become familiar with career opportunities.

Knowing Yourself

Before you can start making career decisions, you need to learn as much as you can about yourself. You have interests, abilities, aptitudes, and values that will play important roles in your future career. In fact, how successful you become in a career and how much you enjoy the career depend largely on these personal characteristics. By learning about yourself, you can better focus on careers that will be best suited to you.

Identifying Your Interests

Your interests are all those things you most like to do. Think about your hobbies and your favorite subjects in school. Activities that you enjoy may be a good basis for a career. Do you like to work with your hands? Do you enjoy solving puzzles? Do you like to be outdoors? Fig. 15-4. The answers to these questions and others like them reflect your interests.

Once you know your interests, you can begin to think about how they relate to career possibilities. For example, if you like working with your hands, you may enjoy a career as an assembler. If you also like being outdoors, you may consider a career as a carpenter.

Identifying Your Aptitudes and Abilities

Interests should be an important factor in choosing a career. However, you cannot base a career choice on interest alone. To be successful in a chosen career, you must have the aptitudes and abilities to do the work well. Aptitudes and abilities are very closely related. An **aptitude** is a natural talent for learning a skill. Fig. 15-5. An **ability** is something you have already learned how to do—a skill you have mastered. Fig. 15-6. For example, if you do well in your algebra class, you probably have the ability to solve mathematical equations. This may also indicate that you have an aptitude for math-related studies. You may find that you have a natural talent for learning math-related subjects.

Ability is often a good indication of aptitude. If you are good at building model airplanes, you probably have an aptitude for working with your hands. If you enjoy experimenting with radio kits, you may have an aptitude for electronics. Make a list of your abilities and aptitudes to refer to when you select a career.

Fig. 15-4. This student likes to work with her hands. What type of career might she be interested in?

Fig. 15-5. An aptitude is a natural talent for learning a skill. This young man has an aptitude for carpentry.

Fig. 15-6. Ann has already perfected the skill of painting. She has the ability to paint a wall.

Identifying Your Values

Your **values** are your beliefs and ideas about things that you think are important. Your values may include security, honesty, family relationships, money, or any combination of these or other values. Fig. 15-7. Take a close look at your values before you choose a career. Suppose you value family relationships. You probably would not be happy in a career in which you were required to work a lot of overtime, even if you were interested in that career. Such a career would disrupt your family life. In this case, you would want to look for a career that would allow you to spend time with your family. When your career is compatible with your values, it can be both rewarding and fulfilling.

For Discussion

As mentioned in the text, your choice of a career will depend on your interests, aptitudes, abilities, and values. Discuss the importance of each of these in coming to a decision about a career.

Fig. 15-7. Tom values people and enjoys spending time with them. What type of career might he find rewarding?

||||| EXPLORING CAREER POSSIBILITIES

To decide which careers interest you, you must first find out what careers are available. Chapter 16 describes some of the careers that are available in the construction field. Although many careers are described in these chapters, many more exist. How can you find out about all the possibilities?

Talking to People

One way to get career information is to talk to people. Parents, relatives, and friends may be able to tell you about their careers. They may also be able to tell you about other types of careers at their places of employment. A school counselor can tell you about many different kinds of careers. He or she may be able to direct you to other sources of career information. Many of these sources contain firsthand information about the duties and requirements of selected careers. Fig. 15-8.

Understanding Dependability

In any job you take, you must be willing to demonstrate dependability. Dependability involves several things. It means, for example, that you will be attentive to the job. It also means that you will carry out a work task thoroughly. Dependability also requires that you be punctual, or on time for work.

You should also understand that good health is important in effective job performance. You should make every effort to avoid foods and practices that can harm your health. As part of being health-conscious, you also should be aware of the advantages of regular exercise. Good health, combined with good grooming, will help make you a more effective employee.

Fig. 15-8. Some companies send representatives to local schools to talk about career opportunities. Ask your guidance counselor about any representatives that are scheduled to appear at your school.

Did You Know?

In medieval times, crafts and skills were sometimes handed down from one generation to the next within one family. For example, a father who was a carpenter might teach carpentry skills to his children. Thus, the same crafts and skills might be practiced in a single family for generations. In fact, family members might become associated with the practice of a certain craft or skill. This might lead others to give them a family name that described their craft. Many family names still used today were given in this way. Examples are Sawyer (one who saws) and Smith (one who works with metals).

Using the Library

Another source of career information is your library. Most libraries have the publications of the U.S. Department of Labor. Two that will be of interest to you are the *Occupational Outlook Handbook* and the *Dictionary of Occupational Titles (DOT)*.

The **Occupational Outlook Handbook** contains information about 200 careers. This handbook refers to careers as *occupations*. The careers, or occupations, are listed in the table of contents under general classifications. Some of the classifications under which construction-related careers may be found are:
- Construction occupations.
- Engineers, surveyors, and architects.
- Technologists and technicians.

When you find a listing that interests you, turn to the correct page and read the entry. The handbook provides information about the duties and responsibilities of each career. It also provides information on the following:

• Working conditions, hours, and earnings.
• Training and education needed.
• The *outlook*, or future, of the career.
• Where to find more information about the career.

This information is updated every two years. The "Index to Occupations" at the back of the handbook lists careers by title. Just after this index is a cross-referenced index to the *Dictionary of Occupational Titles*. The **Dictionary of Occupational Titles (DOT)** describes over 20,000 jobs relating to many different careers. In the *DOT* you can look for construction-related jobs in the index under the name of each job.

Most libraries have other books that describe various careers. Fig. 15-9. You may want to look in the card catalog under "Careers" for books about specific careers. For information on a career you think might be interesting, look in the card catalog under that particular career name.

Magazines, too, can be a source of career information. To find magazine articles about careers, use the *Readers' Guide to Periodical Literature*. In it you will find articles listed alphabetically by subject. It also tells you the magazine in which you can find the article.

Participating in Student Laboratory Enterprises

You can get a general idea of what some construction careers might be like by participating in your laboratory activities. Fig. 15-10. Use this opportunity to learn as much as you can about the duties and responsibilities related to your construction enterprise. Keep in mind, however, that your laboratory experience cannot show you all the aspects of any career. If you decide that you are interested, look for further information about a career in construction.

Taking a Part-time Job

When you reach the minimum employment age, you may want to take a part-time job in a field that interests you. Fig. 15-11. This is an excellent way to become familiar with the requirements and responsibilities of a job. Keep in mind that jobs in many construction occupations

Fig. 15-9. Libraries usually have a good selection of career guidance books.

Fig. 15-10. Your technology education laboratory is a good place to explore your interest in a particular career area.

Fig. 15-11. This student wanted to find out more about a career in construction. He took a part-time job after school as a laborer for a construction company.

require training that you do not yet have. However, you can learn much from simply watching people who are working in jobs you find interesting.

You might not actually do the work you find interesting. However, you would be in a good position to learn more about it. By talking with the other employees, you would be able to learn about advancement opportunities. You might also learn about the advantages and disadvantages of a certain career.

For Discussion

This section of the text discussed several ways in which you might explore careers. As you can see, some of the methods involved library research. Others required personal interviews. Others were related to laboratory activities or part-time jobs. Discuss the advantages and disadvantages of each of these methods.

▌▌▌ GETTING THE RIGHT EDUCATION

Once you have chosen a career in which you are interested, you will need to plan your education for that career. It is wise to plan your career and education in advance. Planning for a career does not mean that you will be "stuck" in a particular career path. If, in preparing for a career, you find that the career does not suit you, you can always change. In fact, the planning you do now will probably help you no matter what career you eventually decide to pursue.

Everyone needs a good education to be able to work well in construction or any other type of career. At one time, graduation from high school was not necessary to get a job in construction. Today, there are very few construction jobs that do not require at least a high school diploma. Employers like to hire workers who are willing to complete whatever they start.

General Preparations

General education is made up of the basic courses, such as reading, writing, mathematics, science, and history, that are required in school. General education is the foundation for further learning. You would have difficulty learning specific career-related skills without first getting a good background in general education. Fig. 15-12.

Fig. 15-12. Construction workers must be able to read notes and dimensions on a set of plans in order to build a structure.

In preparing for a career, you must consider the education you will need in addition to your general education. For example, suppose you are interested in a career that requires many years of college. You may wish to find out now which colleges offer the type of degree you will need in your chosen career. What are the entrance requirements of these colleges? Are your grades good enough to qualify you for acceptance? Good grades now will increase your chances of being accepted into the college of your choice. Also, the good study habits you create now will influence your success throughout your education.

If you are interested in a career that requires apprenticeship or other on-the-job training, you may wish to become familiar with the types of opportunities available in your area. Keep up with developments and new programs that may be offered by private companies. When the time comes to enter such a program, you will be well informed. This will help you make an intelligent decision about which opportunity to take.

Educational Options

The type of education you will need will depend on the career in which you are interested. For example, the training needed to be a carpenter is obviously different from that needed to design machines and equipment. To meet these various needs, many kinds of education and training programs are available.

On-the-Job Training

Some industries train workers on the job. **On-the-job training** is training that a person receives after he or she has been hired. A company may hire a person who has a solid basic education and a good attitude and who is willing to work hard. This person is called a *trainee*. He or she is trained by an experienced worker. The employer may also offer company training sessions to provide trainees with knowledge and skills required on a job. Fig. 15-13. Training may last for a few weeks or for two years or more, depending on the job. As trainees become more skilled and experienced, they advance in their career path. This is an excellent way to earn money while learning the skills necessary for a career.

Apprenticeship

One way to begin a career in most of the skilled crafts is through apprentice training. Fig. 15-14.

Fig. 15-13. Employers sometimes offer workshops to help trainees learn the skills they need.

Fig. 15-14. The construction industry trains many of its workers through apprenticeships.

Fig. 15-15. Apprentices spend a lot of time learning from skilled workers, both in the classroom and on the job.

Plumbers and electricians are examples of people skilled in crafts related to construction. An **apprentice** learns from a skilled worker while on the job. The apprentice also receives classroom instruction. He or she signs an agreement to work and learn for a period of two to five years. Fig. 15-15.

Did You Know?

The practice of training workers through an apprenticeship was known in ancient Egypt. This system of worker training was also used in the Middle Ages in Europe. At that time, the skilled members of those practicing a single craft, such as woodworking, organized into a guild. The experienced craftsmen in the guild were known as master craftsmen. These master craftsmen took on apprentices.

These apprentices were then to learn the skills of the master craftsmen. This period of on-the-job training usually lasted seven years. At the end of that time, the apprentice was expected to demonstrate a skill. The apprentice did this by producing an item that exhibited the skills associated with the craft. This piece was called the masterpiece. This is the origin of a word that is still used.

Technical Institutes and Community Colleges

Some construction-related careers may require only two years of schooling beyond high school. One source of training is a **technical institute**. A technical institute is a school that offers technical training for specific careers. In two-year technical programs, only technical courses are offered. These programs usually include hands-

Fig. 15-16. Community colleges often offer applied technology courses. This class is learning computer-aided design techniques.

Fig. 15-17. This college student is studying for a degree in engineering.

on training to prepare students to obtain a job immediately after they complete the program. Taking courses of this type can help you determine which careers you might like to pursue.

Another source of technical education is a community college. A **community college** is a local, two-year school that is usually supported in part by the state and/or local government. Many community colleges offer courses in accounting, finance, management theory, marketing, industrial supervision, and labor relations. They also offer some technical courses. Figs. 15-16 and 15-17.

Engineers and architects must have at least a **bachelor's degree**. This is a degree awarded for completion of work at a four-year college. They must also have a special state license to practice their profession. Fig. 15-18. Some positions require a higher degree, such as a master's degree or a doctorate. These degrees are awarded for more advanced work beyond that done for a bachelor's degree. Masters and doctorate programs do not usually include any general education at all. Instead, they concentrate on specialized education.

Fig. 15-18. Architects must have at least a bachelor's degree and a state license to practice their profession.

Did You Know?

You may have identified a career. However, you may wonder how you will be able to pay for the education or training you will need. For advice, you should consult your school guidance counselor. Many organizations offer scholarships, loans, and work-study programs. Individual states provide financial aid programs. Also, student loans are available through banks and other private lenders. You should be able to identify the various types of financial institutions. A *commercial bank* offers the widest variety of loans. Loans are also available from a *savings and loan institution*. Savings and loan institutions, however, often specialize in making loans for home mortgages.

The Federal government also provides several kinds of financial assistance to students. Financial help may make the difference in helping you to pursue the career of your choice.

For Discussion

In this section, various methods of obtaining training have been discussed. Among these are technical colleges and community colleges. Are there any technical institutes or community colleges in your town? If there are, do they offer courses that would help you prepare for a career in which you might be interested?

Construction Facts

LIKE FATHER, LIKE DAUGHTER?

Stacey Linsworth is a student at Brindall Junior High School. She does well in most of her school courses, but she is more interested in her father's custom carpentry business.

Recently, Stacey's father asked her if she had thought about what career she wanted to pursue. Stacey told her father that she had always just assumed she would follow in his footsteps and become a carpenter. Stacey's father told her, though, that she should investigate other careers, too.

Stacey decided that she could at least look into a few different types of careers. Her father told her she should think about a career that related to her interests and abilities. Stacey visited her guidance counselor at school and glanced through some of the books the counselor suggested. She discovered many options that she had not even thought about. She decided to begin by doing research on architects and interior designers. She discovered that to work as an architect she would need a five-year college degree as well as a state license.

To become an interior designer, she would have to attend a professional school for three years or earn a bachelor's degree.

When Stacey told her father about her research, he complimented her on her work. He explained to Stacey that the more she learns about possible careers, the better prepared she will be to make a career decision.

Stacey is not yet ready to make a career decision. She is still interested in her father's carpentry business. However, she is becoming more interested in architecture and design. Now Stacey reads everything she can find about these subjects. She is also beginning to take an interest in her grades. After all, she may go to college, and good grades can help her. Right now, she is happy knowing that her preparations will help her when she does choose a career.

CHAPTER **15**

R E V I E W

Chapter Summary

A career is a sequence of related jobs that a person holds throughout his or her working life. Career choices need to be carefully made. In making a career choice, you should consider your interests, abilities, aptitudes, and values. An interest relates to something that you like to do. An aptitude is a natural talent for learning a skill. An ability is a skill you have mastered. Values are beliefs and ideas about things that you think are important. There are several methods of exploring career possibilities. These include personal interviews, using library resources, student laboratory enterprises, and taking a part-time job. In preparing for a career, you need to plan your education. You will need to plan your general classroom education. You might also consider on-the-job training and apprenticeship courses. Technical colleges and community colleges also offer programs of technical education.

Test Your Knowledge

1. Explain the difference between a *job* and a *career*.
2. What are four things you should know about yourself before you choose a career?
3. Why should you consider your values when you make a career choice?
4. What is the difference between an ability and an aptitude?
5. List four ways in which you can get information about careers.
6. Name two publications of the U.S. Department of Labor that can help you learn about career choices.
7. If you wanted to look up information in a magazine about a career, what reference source would you use to locate an article about that career?
8. How can taking a part-time job in a field that interests you help you learn about that field?
9. Name at least four options that will advance your education beyond high school.
10. What is the title given a person who is being trained on the job by an experienced worker?

Activities

1. Look up a construction-related occupation of your choice in the *Occupational Outlook Handbook*. Give the class a brief summary of the information you found.

2. In the library, find a source of career information that is not mentioned in this chapter. Describe the source to the class and the kind of information you found in it.

3. You should have some understanding of personal loans. Your ability to obtain a loan is based on your credit worthiness. Your credit worthiness reflects your ability to repay a loan. Several factors influence your credit worthiness. These include your job, your income, and your past record of repaying your debts. Obtain a blank loan application from a bank. Identify those questions on the loan application that directly relate to the applicant's credit worthiness.

4. Good money management is essential in any career. Once you get a job, the first thing you may want to do is to open a checking account. There are two general types of checking accounts. In one, the checking account holder is charged a certain amount for every check that he or she writes. In the other, the person is charged a flat fee, regardless of the number of checks that he or she writes. Obtain from a bank the basic information on their checking accounts. Evaluate this information. Then prepare a short written report on the advantages and disadvantages of each type of checking account.

CHAPTER **16** **CAREERS IN CONSTRUCTION**

Terms to Know

architect	estimator
bricklayer	ironworkers
building trades	laborers
and crafts	office personnel
carpenters	operating engineers
construction	pipefitters
managers	plumbers
electricians	surveyors
engineer	technicians

Objectives

When you have finished reading this chapter, you should be able to do the following:

- Name some common construction-related trades and crafts.
- Identify some construction-related professions.
- Find more information about specific construction-related careers.

Many people in different occupations work together as a team to create the communities in which we live. Construction projects create many career opportunities all over the United States. As a matter of fact, almost five million men and women are now employed in the construction industry. The figure is still growing. Men and women with a wide range of skills are needed to work in construction. Fig. 16-1.

The development of a building project does not happen overnight. Project development is a big undertaking that requires the skills and talents of many people working together on a team. The team members have the same goal: to complete the building project. Each person does his or her part to achieve this goal.

Workers involved in a construction project range from highly trained people with college degrees to unskilled laborers. Fig. 16-2. Each person plays an important part in the construction process. Construction careers may be classified into three main categories:
- Trades and crafts.
- Construction-related professions.
- Design and engineering professions.

Within each of these categories lies a wide variety of career choices.

TRADES AND CRAFTS

The people who do the physical work of construction are very important. Without their skills and knowledge, a project could not be built. These people work in careers known as the **building trades and crafts**. They are responsible for doing the on-site work needed to complete the project. Most tradespeople learn their trade through a formal apprenticeship program and on-the-job experience. Fig. 16-3. Many tradespeople start as laborers and work their way up to skilled positions. Some vocational and technical schools also offer courses in some building trades, such as carpentry and electricity.

Fig. 16-1. The many people who work at a construction site do many different types of work.

Fig. 16-2. The careers available in the construction industry include a wide range of opportunities. Architects, estimators, laborers, and many other people are needed to complete a project.

Fig. 16-3. Tradespeople learn the skills they need in apprenticeship classes.

Although each type of trade or craft involves different skills and knowledge, most require the same types of qualifications. Tradespeople must be in good physical condition. Their work usually involves lifting heavy objects. Most trades require some knowledge of general math. Nearly all tradespeople must have a good sense of balance. They must be able to work well with their hands.

Workers from many different trades and crafts work together at a construction site. Each trade specializes in a single aspect of the construction work. A few of the most common trades are listed in the paragraphs that follow.

Did You Know ?

The introduction of new building materials has changed the role of the carpenter in building construction. In earlier times, buildings in many parts of the world were built entirely, or mostly, of wood. On such buildings, the carpenter was the principal worker. Now, however, the use of materials other than wood has changed the work role of the carpenter. Today, the carpenter works as one member of a team. Formerly, the carpenter might have been responsible for most of the work on a job. Now, he or she shares the job responsibility with others.

Carpenters

Carpenters are tradespeople who work with wood and wood products. Fig. 16-4. They build the framework for houses and other structures. They also install doors and windows and do any other carpentry that is required. In most homes, carpenters install the kitchen cabinets, wood paneling, and flooring. In heavy construction, carpenters build wooden bridges, piers, and temporary supports for tunnels and bridges. During the construction process, carpenters also build scaffolding to support workers and the wooden forms into which concrete is poured.

The duties of a carpenter vary. They depend on the construction company, the type of construction project, and the preference and skills of the carpenter. Some carpenters do general construction carpentry. Others specialize in one type of carpentry, such as installing cabinets or building bridges and bridge supports.

Fig. 16-4. Carpenters work with wood and wood products.

Bricklayers

A **bricklayer** is a tradesperson who works with masonry. Bricklayers use bricks or concrete blocks and mortar to build walls, fireplaces, floors, partitions, and other structural elements. Fig. 16-5. One special kind of bricklayer is a *refractory bricklayer*. Refractory bricklayers install and repair firebrick linings and refractory tile in industrial furnaces and other high-temperature industrial areas.

Electricians

Electricians assemble, install, and maintain electrical wiring and fixtures. Fig. 16-6. They install wiring in new structures and rewire older structures. An electrician's work includes measuring and cutting lengths of conduit, wire, and cable. Electricians need to be good in math. They must enjoy problem-solving, because each electrical circuit must be calculated accurately. Electricity is dangerous. It is important for electricians to follow safety rules.

Plumbers and Pipefitters

Plumbing and pipefitting are so closely related that they are sometimes considered a single trade. **Plumbers** install, maintain, and repair the plumbing systems that carry fresh water to buildings and wastewater away from buildings. They also build drainage pipes and gas systems in commercial and industrial buildings and in homes.

Pipefitters build and repair pressurized pipes to carry compressed air and steam in HVAC systems and for special applications in factories. They also build special pipe systems for transportation and for power plants. Fig. 16-7.

Fig. 16-5. This bricklayer's skill will determine the appearance of the finished wall.

Fig. 16-6. This electrician is installing a switch.

Fig. 16-7. Plumbers and pipefitters install and maintain piping systems.

Ironworkers

The tradespeople who build steel-framed structures are called **ironworkers**. Fig. 16-8. Ironworkers frequently work high in the air while they weld or bolt the structural steel framework for skyscrapers and bridges. Ironworkers also build and install steel stairs, lampposts, iron fences, ladders, and metal cabinets in industrial and commercial buildings. In addition, ironworkers often weld or bolt prefabricated panels of aluminum or other metal to the sides of buildings.

On the construction site, ironworkers assemble the cranes and derricks that are used to move heavy equipment and materials into place. For example, structural steel, large buckets of wet concrete, and sometimes even equipment must be moved by cranes. Ironworkers are also responsible for positioning reinforcing bars in concrete forms before the concrete is poured.

Operating Engineers

The people who run construction equipment are known as **operating engineers**. Driving construction equipment may sound easy. However, much knowledge, coordination, and skill are needed. Fig. 16-9. For example, crane operators must place large, heavy materials and equipment accurately on the upper stories of tall buildings. Operating engineers also operate bulldozers, excavators, and other kinds of equipment. These workers may go to special training sessions periodically. The sessions are usually sponsored by the equipment manufacturers. The manufacturers teach the operating engineers how to operate and maintain their machines.

Fig. 16-8. Ironworkers sometimes work high in the air to build the structural framework for a building.

Laborers

Laborers do the supportive physical work at the construction site. Although laborers are not considered tradespeople, they are mentioned here because many laborers go on to become tradespeople. The job requires just what its title implies: hard work. Laborers work as assistants to tradespeople. Fig. 16-10. They dig, shovel, clean up, and do other jobs. They also operate motorized lifts and other equipment.

Most beginning jobs for laborers do not require training. Laborers do need to be in excellent physical condition because of the strenuous nature of the job. Laborers do a lot of standing, walking, and climbing. They must also be able to lift heavy objects. Some employers require laborers to have at least a general knowledge of construction methods and materials. A high school education is helpful but not necessary.

Fig. 16-9. Operating engineers need to know how to run large pieces of equipment safely. This crane operator must consider the wind and many other factors to be able to place this precast support accurately.

Fig. 16-10. This laborer is assisting a bricklayer by bringing mortar.

For Discussion

This section of the text has presented information on several different jobs in construction. As you can see, each job is different. To some degree, there have been specialized jobs in construction for centuries. For example, in medieval times, there were guilds. These guilds were organizations of workers skilled in a particular craft. What are the advantages of job specialization in construction?

CONSTRUCTION-RELATED PROFESSIONS

Construction-related professions are those construction-related careers that require specialized, formal education beyond high school. Most of the managerial and supervisory positions fall into this category. Other construction-related professions include support personnel such as estimators, surveyors, and technicians.

Construction Managers

For the work on a project to get done, someone needs to plan how to get it done. On large projects, **construction managers** organize all the necessary materials and people. They assign the work to the workers. Managers must control and check on the work that is being done to see that the workers are doing their jobs properly. Managers also determine whether the work is being done according to specifications. Without managers to give orders and to oversee the job, little would be accomplished. Some jobs might have to be done over because they were done improperly or in the wrong order.

Project Managers

The project manager is responsible for the construction project from start to finish. Fig. 16-11. A project manager usually works in the home office. He or she may be responsible for more than one project at a time.

Fig. 16-11. The project manager manages the project. He or she is responsible for checking all aspects of the job.

The project manager plans, organizes, and controls a construction project. He or she must be able to manage and lead people, make decisions, and keep track of several projects at once. Most project managers have college degrees. Project managers are required by many companies to have engineering skills and on-the-job construction experience.

Construction Superintendent

The manager that is directly in charge of one particular project is called the construction superintendent. He or she works under the direction of a project manager. The construction superintendent has basically the same duties as the project manager, but on a smaller scale. The construction superintendent usually has a field office at the site. Fig. 16-12.

The construction superintendent's job starts before the construction begins. He or she develops schedules and decides when each construction task should be started and finished. Then the superintendent hires the workers and checks on materials and equipment. Once the construction begins, the construction superintendent supervises the workers, keeps the job on schedule, and tries to stay within the budget.

Construction superintendents are usually promoted from a construction-related trade or craft. Many have college degrees in construction technology. Construction superintendents need to be able to make good decisions under pressure because deadlines need to be met constantly. Management and leadership skills are also essential.

Other Construction-related Professions

Construction-related professions cover a wide range of careers in addition to managerial positions. Estimators, surveyors, and technicians are examples of professionals that contribute to a construction project. Office personnel such as secretaries and receptionists also contribute to the overall project.

Fig. 16-12. The construction superintendent is responsible for everything that happens on the site.

Estimators

The **estimator** for a project carefully calculates what the job will cost. The contractor depends heavily on the skills of the estimator. The success or failure of a project depends on the estimator's ability. If the estimator's figures are too high, the contractor may not get the job. If the estimator's figures are too low, the contractor cannot make a profit on the project. Without profits, a contractor cannot stay in business. Therefore, a good estimator plays a very important part in the construction industry.

The estimator bases his or her estimate on the cost of materials, labor, equipment, overhead, and the amount of profit that the contractor wants to make. To do this, the estimator must be able to read construction prints and understand specifications. He or she must, of course, have good math skills. Construction experience is also helpful.

Surveyors

Surveyors measure and record the physical features of the construction site. They survey the property and establish its official boundaries. Fig. 16-13. They also lay out, or measure and mark, the construction project. Surveying is an important part of construction and must be done very precisely. If the surveyor makes a mistake, the whole project could be built in the wrong place or to the wrong specifications.

Fig. 16-13. The surveyor determines the precise boundaries of the site.

Technicians

Technicians are construction personnel who work neither in the home office nor at the site. They work in laboratories, testing soil samples and concrete and asphalt materials being used on the project. Fig. 16-14. These workers test the sample that is taken from each batch of concrete that is poured. Technicians work for individual testing laboratories. They are usually trained by the laboratory for which they work.

Office Personnel

Office personnel is a general term for all of the clerical and secretarial employees of a company, together with their managers and super-visors. An *office manager* runs the business office of a construction company and oversees the work of the office personnel. He or she is responsible for contracts and other legal requirements for each project. *Secretaries, clerks*, and *receptionists* perform tasks such as typing, filing, and answering the telephone. Fig. 16-15. Office personnel keep detailed records and provide the information needed by the contractor to keep construction operations running smoothly.

Fig. 16-14. Engineers need to know about the soil conditions on a site before they can design a proper foundation. Here a technician performs a soil test using a machine that brings up soil samples.

Fig. 16-15. The office personnel help keep the company running smoothly.

For Discussion

As you have read, any construction project requires people working behind the scenes. These jobs range from the technical to the routine. If the members of this support team are to work together, they must have certain qualities. Discuss the qualities that the members of the support team should have if they are to provide the greatest help to the construction team.

DESIGN AND ENGINEERING PROFESSIONS

Architects and engineers often work together to design and engineer each element of a project. For example, a shopping center building is designed by an architect. An engineer designs the utility system, drainage plan, and structural system for the shopping center. The architect is concerned with making the overall design. The engineer is concerned with the structural features that the design requires.

Architects

Architects create new building designs. They plan details such as the efficient use of space for homes, office buildings, shopping centers, and community projects. Fig. 16-16. Architects are responsible for the beauty and usefulness of a structure. They must mold the owner's ideas into a workable, problem-free design. As the construction progresses, the contractor may call upon the architect to solve design problems as they arise.

Architects must have a college degree in architecture. Creativity and the ability to make decisions and share ideas are also very important characteristics for an architect. He or she must be able to get along well with people and must be able to express ideas clearly.

Engineers

Engineers are responsible for the structural design of a project. Fig. 16-17. They work with architects to be sure that the architectural design is structurally sound. The engineer's responsibilities include making sure that the structure is safe, that the utility systems are designed properly, and that site drainage is adequate. In other words, the engineer is responsible for all of the details that make a structure sound, safe, and convenient.

Many different kinds of engineers may work together on a large construction project. *Civil engineers* are responsible for correct land elevation and structural layout. *Structural engineers* oversee the placement of structural materials and calculate the load that will be supported by each structural element. *Electrical engineers* oversee the installation of electrical wiring. *Mechanical engineers* assist with the plumbing and piping layouts.

In heavy construction, engineers plan and design bridges, dams, highways, tunnels, and other structures. The amount of stress that each of these structures will receive is a critical factor. For example, a dam must be able to withstand the pressure of the huge wall of water it will hold back. A bridge must withstand the weight of heavy traffic as well as the forces of wind and water. The engineer calculates the stress that each structure must withstand. He or she then designs the structure accordingly.

Fig. 16-16. Architects plan the use of space in a building.

Fig. 16-17. This engineer designed part of this project. Now the engineer is checking the construction.

Did You Know?

The civil engineer is concerned with the building of dams, bridges, and highways. He or she is also involved in the construction of buildings. The first school of civil engineering was founded in Paris in 1747. This school was the National School of Bridges and Highways. Here, students were instructed in the basic principles of engineering. This school was the first successful attempt to set up a specialized program of instruction for civil engineers.

For Discussion

On any construction project, there must be close cooperation between the architect and the engineer. As mentioned, the job of the architect and the job of the engineer are similar. Each job requires special technical knowledge. Good math skills also are essential. Though the jobs of an engineer and an architect are similar, can you think of any special skills that an architect might need?

YOUR CAREER INTERESTS

If you are interested in a career in the construction industry, perhaps you should find out more about those careers that interest you most. One good way to find out about a career is to look it up in a career handbook such as the *Occupational Outlook Handbook* or the *Dictionary of Occupational Titles*. These career guides can tell you what each career involves and what kind of education or training is needed. They can also tell you what the average pay is and whether the need for workers in that field is growing or declining.

Another way to find out about construction-related careers is to visit a building site. Then you can see for yourself what the different jobs involve. You may be able to talk to some of the workers during their breaks to find out more about construction.

If you decide you are interested in one of the construction trades, you may consider taking a part-time job as a laborer. This will enable you to experience construction work firsthand. Remember, you will not be able to start out at the top with high pay. You will probably have to start at the minimum wage and work your way up. If you find that you would rather do a different kind of work, you can. If you decide to stay in a construction-related career, you will have a head start in gaining the necessary knowledge and experience.

THE ENTREPRENEUR AND CONSTRUCTION TECHNOLOGY

A person who starts a business is usually an **entrepreneur**. This term may be unfamiliar to you. However, you certainly know some entrepreneurs. An entrepreneur is anyone who organizes and manages a business. This person also assumes the risks of the business. This means that the entrepreneur is responsible for paying the business expenses. All responsibility for the success of the business rests with the entrepreneur. A person who is self-employed, or in business for himself or herself, is an entrepreneur.

All entrepreneurs face four similar problems. These problems are:

- Identifying a need.
- Finding a product to satisfy the need.
- Financing the business.
- Selling the product.

Perhaps you will decide that you would like to be in business for yourself. Being in business for yourself is not always easy. If you are to be successful, you must be able to deal with each of the four problems listed above.

The spirit of entrepreneurship has been important in construction. Several massive construction projects were guided by the energy and vision of a single individual. For example, the Eiffel Tower and the Suez Canal were among the most remarkable construction projects of the time. Each of these projects was considered risky. Many said they could not be built. Yet, each of these projects was guided to its successful completion by the energy and vision of a single person. That person was an entrepreneur. An entrepreneur is a person who assumes the risks of a business in the hope of making a profit from it. In the case of the Eiffel Tower, the entrepreneur was Gustave Eiffel. In the case of the Suez Canal, it was Ferdinand de Lesseps.

The late nineteenth century saw a rapid growth in entrepreneurship. That growth has continued into our own time. Entrepreneurs today are responsible for much of the growth in the construction industry. Many construction businesses are owned by entrepreneurs, people who invest their money and effort into their own business. A successful entrepreneur is also a leader. He or she is able to inspire and lead a construction team. The practice of effective leadership is essential to entrepreneurship. To be an effective leader, the entrepreneur must also understand teamwork.

In our time, the entrepreneur has attracted the attention of investors known as venture capitalists. These are people who provide capital (money) for someone to use in a business. Generally, venture capitalists invest in new technologies. Of course, these new technologies also attract investment from other individuals. In fact, one of the foremost construction projects of our time, the Chunnel, is being financed in part through the sale of shares of stock to individual investors.

Not everyone is suited for self-employment. Some people are more comfortable and happy working for others. In examining career choices, though, you will want to consider all aspects of construction technology.

Understanding Yourself

In any job you take, you should have a good understanding of your own abilities. Not every aspect of a job may be appealing to you. For example, some parts of the job may be more attractive than others. To be effective, though, you should give equal attention to all aspects of your job.

You should also recognize the importance of learning to accept criticism. If the criticism is deserved, you should be willing to put it to good use. You might, for example, use it to improve your job performance.

Human relationships will be important in your career. You will probably be working with other people. Your success on the job will depend heavily on your ability to get along with the other members of your work team. To get along with others, you will need to exercise self-control and cooperation.

DEVELOPING LEADERSHIP

We live in a democratic society. In our society, everyone is urged to develop their skills to the fullest. By doing this, they will be able to participate as more active members of society. In developing your skills, you also can develop some of the qualities of leadership.

All of us have known leaders. You may have noticed that all leaders have certain characteristics. For example, good leaders are good communicators. They are also able to concentrate their attention on a single project. They are determined. They are able to instill confidence in others. Of course, these qualities are more apparent in some people than in others. However, all of us can develop leadership skills. One way of developing such skills is by joining student clubs. Club membership provides an opportunity to practice effective communication. It also offers you practice in teamwork and the other skills needed to work within a group. Throughout your life, you will be given many opportunities for leadership. Now is the time to develop the skills you will need.

Construction Facts

CHANG CONSTRUCTION COMPANY

Thoun Chang is a student with a very special dream. He wants to own and operate a construction company. Ever since he was a child, Thoun has been fascinated by the construction of large buildings and by the people who build them. Although he is only in junior high, Thoun knows how he will prepare for his career. Even now, he is beginning his preparations by taking a technology education course at his school.

Thoun has talked to his guidance counselor at school, to his parents, and to a neighbor who is a field engineer for a local construction company. From them, he has gained various insights and viewpoints about the construction industry. His neighbor advised him to take business courses in school. He might even want to go to college and major in business. He will need plenty of knowledge and business experience to operate a large company successfully.

At the same time, he will need to know every phase of the construction work. He plans to gain this experience slowly, while still in school. As soon as he can, he will get a part-time job as a laborer in a construction company. He knows he can learn a lot just from being at a construction site. He would like to spend his summers doing various types of construction jobs.

As Thoun prepares for his chosen career, he encourages his friends to "start looking" now to find a career they might enjoy. He thinks that having a purpose makes school more interesting. Thoun knows that he has a long way to go before he accomplishes his goal. However, he is satisfied that he is doing all he can to prepare himself to be the owner of Chang Construction Company.

CHAPTER 16

R E V I E W

Chapter Summary

Construction projects employ people in a variety of trades and occupations. Workers in the building trades and crafts are responsible for doing the on-site work needed to complete the project. These workers include carpenters, bricklayers, electricians, plumbers, pipefitters, ironworkers, operating engineers, and laborers. Workers in construction-related professions often work as managers or supervisors. Other workers are support personnel, such as estimators, surveyors, and technicians. Planning on a construction project is done by construction managers.

The project manager is responsible for the construction project from start to finish. The manager directly in charge of one particular project is called a construction superintendent. Construction-related professions include estimators, surveyors, technicians, and office personnel. Architects create new building designs. Engineers are responsible for the structural design of a project. If you are interested in a career in construction, you should find out more about that career by checking the *Occupational Outlook Handbook* or the *Dictionary of Occupational Titles*. These books will be at your library.

Test Your Knowledge

1. What are the three main categories of construction careers?
2. In what two ways do most tradespeople get their training?
3. Name at least four careers that are considered construction trades.
4. What is the name given to workers who assist tradespeople?
5. What is a construction-related profession?
6. Name two differences between a project manager and a construction superintendent.
7. What professional is responsible for the usefulness of the space in a building?
8. What professional is responsible for structural soundness?
9. Name four different types of engineers that might be involved in a large construction project.
10. Name at least two ways to find out more about a construction-related career.

Activities

 1. Look up a construction-related career of your choice in the *Occupational Outlook Handbook*. Give the class a brief summary of the information you find.

2. Visit a building site near your home. Make a list of all the different types of careers that are represented at the site.

3. At times, there is more construction than at other times. The rise and fall of residential construction follows the economy very closely. On the other hand, because of the amount of lead time needed for large construction projects, commercial and heavy construction may be quite plentiful during economic downturns. This factor also helps minimize employment in the construction industry. Prepare a short written report on the present level of construction activity in your community.

4. You may not be happy in the job you choose. If this is the case, you may want to find a new job. Generally, it is important that you obtain a new job before leaving your old job. This will help you avoid anxiety. You will want to leave your job in good standing. You will want to work hard and attentively through your last work day. It will also be important for you to obtain a good recommendation from your employer. Assume that you are unhappy in your present job. Research procedures for leaving a job. Then prepare a short written report outlining your plan for getting another job in the same field. Your report should include information on ways of identifying possible new employers, securing an interview, and presenting your job qualifications. You also should include information on giving your employer notice that you intend to leave the job. You should arrange for his or her recommendation.

5. Find out how many construction permits were issued in your community last year. Find out also how many construction permits were issued the year before that. Figure out the percentage of increase or decrease in construction activity in your community.

Activity 1: Measuring Height Using an Inclinometer

Objective

An inclinometer is an instrument used to measure height. After completing this activity, you will know how to use an inclinometer to measure the height of a tree, building, power pole, and other objects.

Materials Needed

- 10 sheets of grid paper (10 squares per inch)
- 1 small bench level
- 1-6″ protractor
- 1 squared block of wood 1½″ × 7″ × 5″
- 1 nut for a ¼″ carriage bolt
- 1¼″ carriage bolt, wing nut, and washer
- 3-50″ sections of ¾″ metal conduit
- 1-10″ circle of ⅝″ or ¾″ plywood
- 3-3½″ × ¼″ carriage bolts with washers and wing nuts
- 3- ½″ metal screws
- 1-60″ length of lightweight chain

Steps of Procedure

1. Figure A illustrates the general construction of the inclinometer. The bench level will be used to level the inclinometer from front to back and side to side. It is important to position the protractor on the side of the board so that the 90° line is exactly parallel with the bench level. The accuracy of your inclinometer begins with this placement. A hot-glue gun can be used to attach the protractor to the board.
2. Use a small, sharp knife to carefully cut a hexagon-shaped hole in the bottom of the

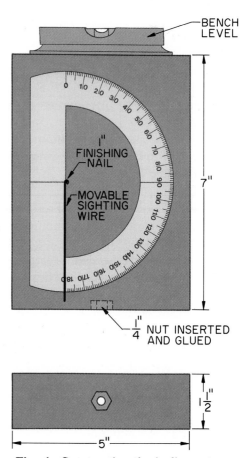

Fig. A. Constructing the inclinometer.

board. This hole should be ¼″ deep. Place a small amount of hot glue on the flat sides of the nut and insert it into the hole. Be careful not to get glue inside the nut, where the threads are located. This nut will be used to attach the inclinometer to a photographic tripod or a stand you can construct.

3. The final step in constructing the inclinometer is to add a sighting pointer. This is accomplished by first placing a small, 1″ finishing

nail at the intersection of the 0°, 90°, and 180° points of the attached protractor. Allow about ½″ of the nail to extend from the board.

4. Select a small, straight wire 5″ in length. Use a pair of small needlenose pliers to form a loop at one end. Place this loop around the nail. Tighten it so it swings freely but does not fall off. Use a wire cutter to shorten the wire so it extends about ¼″ past the protractor's outer edge. Be careful not to bend this wire. It is your pointer to obtain accurate degree readings on the protractor.

5. If you do not have a photographic tripod, you will need to construct a tripod stand. This stand will have several uses in other activities you may develop. Figures B and C show the general construction of this stand.

6. Flatten three 50″ pieces of ¾″ metal conduit on one end. Drill as shown with a ¼″

metal drill. The square ends can be rounded on a grinder. Fig. C.

7. Prepare the wooden attachment board from a round piece of ⅝″ to ¾″ plywood 10″ in diameter. Cut the wings and notches in the plywood disc using a coping saw.

8. Drill each projection for a 3½″ × ¼″ carriage bolt, washer, and wing nut. It is recommended that you attach a stabilizing chain about halfway down the conduit to join all three legs. Fig. C. A ½″ metal screw can be used for this purpose. Drill a ¼″ hole in the center of the attachment board for attaching the inclinometer.

Fig. C. Constructing the tripod.

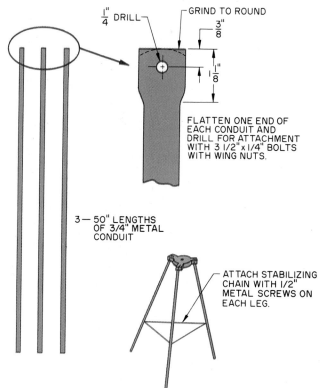

Fig. B. Constructing the wooden attachment board.

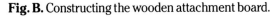

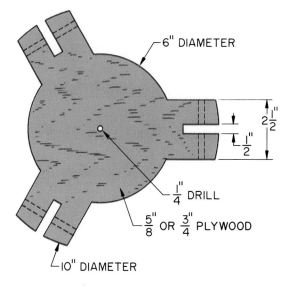

ACTIVITIES

9. Mount the inclinometer on a tripod stand. Select a tree and move far enough away so you can easily see both the base and top of the tree. Position the inclinometer with the 90° edge of the protractor pointing at the tree. Carefully adjust the legs to bring the inclinometer into level from front to back and from side to side.

10. Next, use a tape measure to determine the exact distance from the base of the tree to the inclinometer. For final graphing of the data it is best if the selected distance is 25′, 50′, 75′, or 100′. Also measure the distance from the ground to the 90° line on the protractor. This may be useful depending on how your teacher wants you to analyze your data. Fig. D.

11. You are now ready to make your first measurement. This is best done with a partner. Use your finger or a pencil point to carefully raise the sighting wire until it is pointing at the base of the tree. Sight along the wire. Have your partner record the degree reading. If you are on level ground, the degree reading will be slightly more than 90°. If you are uphill from the tree, it will be considerably more than 90°. If you are downhill from the tree it will be less than 90°.

12. Make your final reading by raising the sighting wire to point at the top of the tree. Fig. D. Again, have your partner record the degree of angle. Repeat each measurement two or three times to confirm your readings.

Fig. D. Measuring the height of a tree.

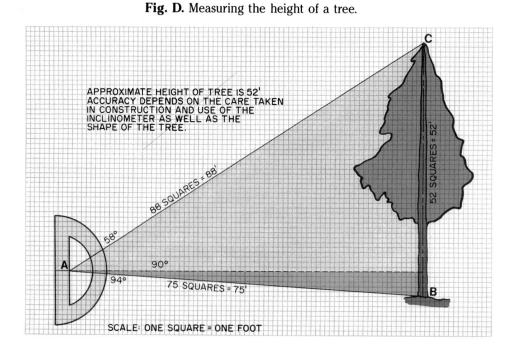

ACTIVITIES

13. The data you have collected can now be graphed to show the height of the tree. The data collected for Tree "A" follows:

DATA FOR TREE "A"

DISTANCE FROM TREE BASE	75′
DEGREE READING OF TREE BASE	94°
DEGREE READING OF TREE TOP	58°

Locate a point on the grid paper near the left-hand side as shown in Figure D. This will be Point A. Place the protractor on the point as indicated and mark the 58° and 94° positions. Use a straightedge and pencil to join and extend lines through Points A and B and Points A and C.

14. Place a piece of grid paper along line AB. Count over 75 squares from Point A. Since there are 10 squares in each inch this is a distance of 7½″. Each square represents one foot. Now use a straightedge and pencil to extend a line from Point B to intersect with line AC. This intersect point becomes Point C.

15. Now count the number of squares between Point B and C. Depending on the accuracy of your construction, the count should be near 52 squares. Thus, the tree is approximately 52′ in height.

16. Refer to Fig. E. Tree "B" was measured in a similar manner. Note that the 90° line points to a position about 18.5 squares or 18.5′ up the tree. It is assumed this measurement was taken while standing uphill from the tree. Also note how a strip of grid paper is placed along lines AB and AC to determine the distance.

Fig. E. Graphing the results.

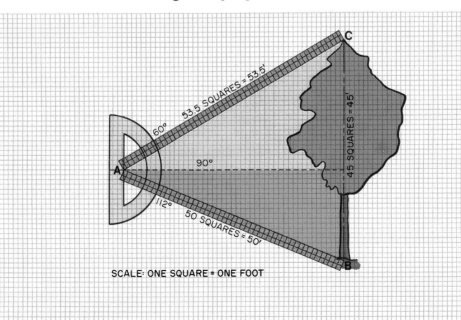

Activity 2: Conducting a Soil Compaction Test

Objective

After completing this activity, you will know how to construct and use a simple device to measure soil compaction. This is an important test prior to beginning a construction project.

Materials Needed

- 2 blocks of wood 1½″ × 4″ × 9″
- 1-1″ × 4″ × 10′ length of lumber
- 3 weights, 5 lbs each
- 1-8″ × 12″ piece of ½″ plywood
- 1-12″ ruler
- 1 watch with a second hand

Steps of Procedure

1. Each student will construct a simple, four-sided box frame from 1″ × 4″ × 10′ lumber. Construct the frame shown in Fig. A. This frame should have an inside dimension exactly 4″ wide × 9″ long × 4″ deep. The surface area of the frame is 36 square inches. The box has a cubic volume of 144 cubic inches.

2. The 1½″ × 4″ × 9″ solid blocks of wood may be cut from a piece of 2″ × 6″ stock lumber. These blocks of wood should fit loosely inside the frame constructed above. Be certain they do not bind when slid down in the frame. These blocks will serve as pistons when the soil sample is compacted.

3. Weigh the blocks and record their weight.

4. Each student or team of students should bring soil samples to test. The size of the sample should be large enough to fill the

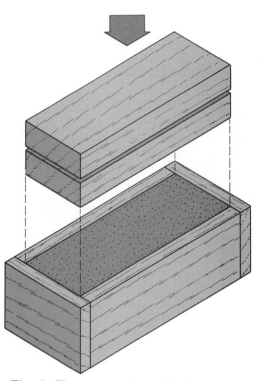

Fig. A. The construction of the box frame.

wood frame level full. A wide variety of samples will prompt comparison and discussion of the characteristics of the different samples.

5. Place the wooden frame on the 8″ × 12″ piece of ½″ plywood. Pour the soil sample into the frame. Using a metal straightedge, scrape away the excess soil to make it exactly level with the top of the frame. Be careful not to compact the soil while doing this.

6. Now place the two 1½″ × 4″ × 9″ blocks of wood on top of the soil sample so one block is exactly on top of the other. Immediately measure and record the exact height

the blocks extend above the upper edge of the wooden frame. Place a 5-lb weight on top of the wood blocks.

7. At one-minute intervals, measure and record the height the two blocks extend above the wood frame. Continue this measurement until the wooden piston stops moving or until you have made at least 15 measurements. Fig. B.

8. The amount of load present on each square inch of the soil's surface is easily calculated. The block has a surface area of 36 square inches. Divide 5 lbs by 36. The result is the pressure on each square inch of surface. You will discover this is 0.1388 lbs per square inch. Divide the combined weight of the two blocks of wood by 36 and add the result-

ing answer to 0.1388 lbs to obtain the total weight on each square inch. Assume the blocks have a combined weight of 1.2 lbs. This would add 0.0333 lbs per square inch, giving a total of 0.1721 lbs per square inch.

9. Use the data you have collected to create a simple line graph showing the rate (speed) and amount (distance) of compaction for the soil sample. Fig. C.

Application

The degree of soil compactibility is very important when constructing a shed, house, building, or other large structures. It may be necessary to check the compactibility characteristics to a depth of several feet. Soil compactibility differences help to explain much of the damage or lack of damage in the various areas of San Francisco during the 1989 earthquake.

Fig. B. Recording the settlement values.

INCHES	MINUTES
3.75	1
2.25	2
1.25	3
0.75	4
0.65	5
0.50	6
0.50	7
0.50	8
0.50	9
0.50	10

Fig. C. Plotting the settlement values.

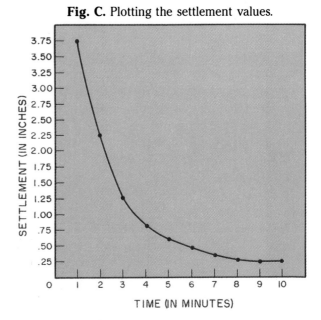

ACTIVITIES

Activity 3: Measuring Distance with a Plane Table

Objective

After completing this activity, you will know how to use a simple plane table to create maps similar to those developed by early mapmakers. You also will have developed practice in the use of the metric system of measurement.

Materials Needed

- paper
- 2-6″ protractors, each 152 mm wide
- 1 small bench level
- 1 squared block of wood 38 mm × 178 mm × 127 mm
- 1 nut for ¼″ carriage bolt
- 2 sewing needles at least 35 mm long
- 1 piece of wood 5 mm × 8 mm × 83 mm
- metric scale 300 mm in length
- metric stick or metric tape measure

Steps of Procedure

1. This activity is a metric activity. It will help you gain practice in the use of the metric system of measurement. Though the United States has used the customary system of measurement, most other countries use the metric system of measurement (also called the International System of Units).

   In the customary system, measurements of length are made in feet, inches, or parts of an inch. In the metric system, measurements of length are made in kilometers, meters, and centimeters. The way in which these units relate to one another is given in Table A. Table B lists some metric equivalents of customary units.

2. Your teacher may provide materials not listed above to complete this construction.

3. The construction of the plane table is similar to the construction of the inclinometer described in Activity 1, Section V. It will be necessary to use a tripod to support the plane table. If you do not have access to a camera tripod, you can construct a tripod identical to the one described in Activity 1, Section V.

4. Figure A illustrates the general construction of the plane table. Attach a 152-mm plastic or metal protractor on one side of the 127-mm × 178-mm wooden block. A hot glue gun can be used to make this attachment. Place the straight side of the protractor exactly parallel to the long edge of the wooden block.

5. Place the ¼″ nut in the center of the lower surface of the 127-mm × 178-mm block. Use a small, sharp knife to cut a hexagonal hole 6 mm deep in the bottom of the board.

Table A. Metric Units of Linear Measure

Property	Unit name	Symbol	Relationship of units
LINEAR MEASURE	millimeter	mm	1 mm = 0.001 m
	centimeter	cm	1 cm = 10 mm
	decimeter	dm	1 dm = 10 cm or 100 mm
	meter	m	1 m = 100 cm or 1000 mm
	kilometer	km	1 km = 1000 m

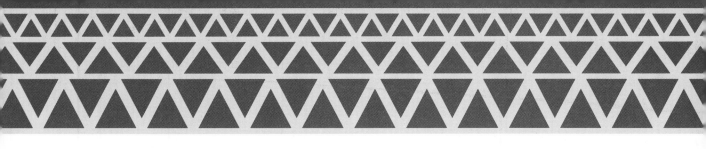

ACTIVITIES

Table B. Metric Equivalents of Some Customary Measures of Length

Customary Measure	Metric Equivalent
1 inch	2.54 mm
1 foot	0.304 m
1 yard	0.914 m
1 mile	1.60 km

Then place a small amount of hot glue on the flat sides of the nut and insert it into the hole. Be careful not to get glue on the inside area where the threads are located. This nut will be used to attach the plane table to the tripod.

6. Construct the pointer as shown in Fig. A. After shaping the pointer, make a hole 5 mm from the square end of the pointer exactly in the middle of its 8 mm width.

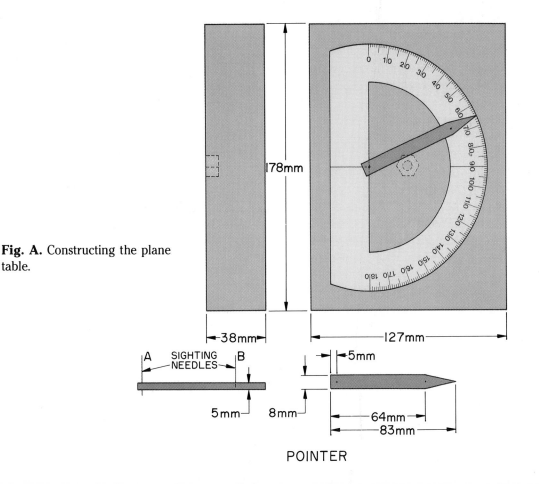

Fig. A. Constructing the plane table.

178mm

38mm

127mm

A   SIGHTING NEEDLES   B

5mm

5mm   8mm

64mm

83mm

POINTER

7. Place the second needle 64 mm from the first needle's location and exactly in the middle of the 8-mm width about two-thirds of the way through the pointer. When sighting along the pointer, the two needles should align with each other and the pointed end of the wooden sighting bar. The needles should be exactly straight up-and-down.

8. You are now ready to collect mapping data. To complete this activity, it is best to work in teams of two. Select an area along a straight section of sidewalk or fence. Use a metric stick or metric tape measure to locate two points exactly 15 m apart. Mark these two points with a piece of chalk or short wooden sticks. As you face the area you are to measure, let the point you marked on the left represent Point A and the other Point B.

9. Place the tripod over Point A so the middle of the protractor is exactly over that point. The long edge of the plane table should be parallel to the area you are to measure. To test this, move your pointer to the 180° mark. The pointer should be pointing exactly at Point B. If it is not, keep turning the table until it does.

10. Use the bench level to level the plane table from left to right and front to back.

11. Once you have positioned the plane table, you are ready to begin collecting data. It is possible to take several sightings of various objects before moving to Point B, providing you can also see the same objects from Point B. You may want to check this before beginning the readings. Also, it is best to sight on objects that are no more than 15 m to 25 m from your table.

12. Now select the first object to sight from Point A. Record a brief description of the object and assign it No. 1. Sight along the wooden pointer. Move the pointer until sighting needle "B" is in perfect line with the stationary sighting needle "A." Record the degree reading at the end of the wooden pointer on your chart under "Readings for Point A."

13. Repeat this process for other objects you would like to sight from Point A. Number these objects No. 2, No. 3, etc.

14. Move your plane table to Point B. Once again, level and align it for sightings to the objects you viewed from Point A. This time, check your alignment with Point A by placing the wooden pointer on 0° and turning the table until the wood pointer is pointing exactly at Point A. Now move the pointer to point at each object viewed from Point A. Record the readings on your chart. Once you have completed this series of readings, you are ready to place your data on the paper. **NOTE:** Do not use your plane table to measure Point A and your partner's table to measure Point B. Being hand constructed, they will have minor differences and produce greater error in the data. Use the same instrument at both points.

15. Use a straightedge to draw a line 100 mm long across the bottom edge of your paper. The left end of the line will represent Point A. The right end, a distance of 15 m away, will represent Point B. Now use the protractor at Point A to project the first degree reading as shown in Fig. B. Use a straightedge to extend a line from Point A through the mark. Now repeat this process at Point B with the reading to the same object. Use

the straightedge to extend the line until the two lines intersect at Point C.

16. Use the metric scale to measure the distance to the object. One centimeter will represent a distance of one meter. Each millimeter will represent one decimeter. Use the 300-mm scale to measure the distance along line AC and BC. Write both the degree readings and the number of meters, decimeters, and centimeters from one point to the other. See Fig. B.

17. Are you ready to test the accuracy of your instrument? If so, go back to Point A with your partner and the meter stick or metric tape measure. Measure the exact number of meters and parts of a meter from Point A to the object you have just recorded on your paper. Then go to Point B and also measure the distance. These measurements should not vary more than a few centimeters from what you plotted on the paper. **NOTE:** This will be true only if the land is flat and you have carefully constructed and used your plane table. You are sighting in a straight line. In hilly country you may have to walk down 100 meters and up 100 meters to get an object only 35 meters from your plane table.

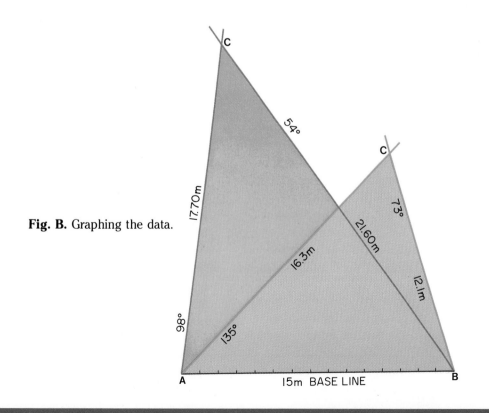

Fig. B. Graphing the data.

▎▎▎GLOSSARY

A list of words in construction technology, with their meanings

ability. Something you have already learned how to do—a skill you have mastered.

accident. An unexpected happening that results in injury, loss, or damage.

adhesives. Materials that hold, or bond, other materials together.

admixtures. Anything added to a batch of concrete other than cement, water, and aggregate.

aggregate. The sand and rocks used in concrete.

alterations. Changes in the structural form of a building or other structure. These changes are usually made to make the structure more attractive or useful.

apprentice. One who learns from a skilled worker while on the job. The apprentice also receives classroom instruction.

aptitude. A natural talent for learning a skill.

arch bridge. This type of bridge uses an arch to carry the weight of the bridge. Arch bridges are made of concrete or steel and are usually constructed over deep ravines.

architects. People who create new building designs. They plan details such as the efficient use of space for homes, office buildings, shopping centers, and community projects.

architectural drawings. Drawings that show the layout of a building. Some of the most common kinds of architectural drawings are floor plans, site plans, elevations, section drawings, and detail drawings.

arc welding machines. A machine used to weld materials such as steel beams at construction sites. An arc welding machine, or arc welder, uses an electric arc to melt portions of the metal beams and thus weld them together. A gasoline engine powers the machine. The engine turns an electrical generator, which provides the electric arc.

armored cable. Cable made of several wires inside a flexible metal casing. It is used in dry environments where sturdy, yet flexible, wiring is needed.

asphalt. A petroleum product made from crude oil.

bachelor's degree. This is a degree awarded for completion of work at a four-year college.

backhoe. A type of excavator that is used for general digging. It is usually mounted on either a crawler or a truck frame. A dipper bucket is attached to a boom that is operated by hydraulic cylinders. The bucket is designed to dig toward the machine.

backsaw. A handsaw with a very thin blade reinforced with a heavy metal back. This keeps the thin saw blade from bending. The backsaw is used to make very straight cuts, such as those on trim and molding.

bar chart. A chart that is easy to read and interpret. For example, months may be listed across the top of the chart. All the major jobs are listed down the side of the chart. A bar is then used to show the starting and completion dates for each job.

batter boards. Boards held horizontally by stakes driven into the ground to mark the boundaries of a building. String is used to connect a batter board with another at the opposite end of a wall. The batter boards are placed outside the building's boundaries so that the attached strings cross over the corners of the building.

bearing-wall structure. In this type of structure, heavy walls support the weight of the building. This type of structure has no frame.

bids. A company quotes a price for which it will do a particular job. Construction companies obtain most of their work through the competitive bidding process.

blind riveter. A tool used by sheet metal workers to fasten pieces of sheet metal together. With a blind riveter the worker can do the entire riveting operation from one side of the sheet metal.

board lumber. Lumber that measures less than 1½ inches thick and 4 or more inches wide.

bond. A bond provides protection for the owner in the event the contractor does not follow the terms of the contract. To get a bond, a construction company must have a good reputation and financial dependability. The contractor pays a fee to the bonding company. In return for the fee, the bonding company issues the bond.

boom. The long arm of the crane that directs the cable.

bricklayer. A tradesperson who works with masonry. Bricklayers use bricks or concrete blocks and mortar to build walls, fireplaces, floors, partitions, and other structural elements.

brick trowel. Masons use brick trowels to place and trim mortar between bricks or concrete blocks. Brick trowels are usually made of steel and have handles of wood or sturdy, high-impact plastic.

building trades and crafts. The skills practiced by people who are responsible for doing the on-site work needed to complete the project.

built environment. Structures such as buildings, bridges, and highways as well as other parts of the environment that people have shaped or altered.

bulldozer. A tractor equipped with a front-mounted pushing blade. The bulldozer is one of the most basic and versatile pieces of construction equipment. One of its primary parts is the blade, which is attached to the frame of the machine. It is designed for clearing land of bushes and trees.

bull float. A tool used by cement finishers to smooth the surface of wet concrete. The face of the float, the part that touches the cement, is made of softwood.

cantilever bridge. This type of bridge is used for fairly long spans. It has two beams, or cantilevers, that extend from the ends of the bridge. They are joined in the middle by a connecting section called a suspended span. The whole structure usually receives additional support from steel trusses.

career. A sequence of related jobs that a person holds throughout his or her working life.

carpenters. Tradespeople who work with wood and wood products.

certificate of occupancy. A certificate issued by the building inspection department which approves a building for use. This certificate shows that the building has passed the building inspector's check of the structure.

chalk line. A chalk line, or chalk box, is used to mark a straight line. A chalk line consists of a line (string) that is contained in a housing filled with chalk. The chalk coats the line as the line is drawn out of the housing.

claim. A legal demand for money.

claw hammer. A common type of hammer. The face, or pounding surface, of the claw hammer is used to drive nails. Opposite the face is a V-shaped notch called a claw. The claw is used to remove nails from boards.

cofferdam. A watertight wall, built to keep water out of the worker's way. This temporary wall can be made of timber, concrete, soil, or sheets of steel.

cold chisel. Cold chisels are made of solid steel. They can be used to cut sheet metal, round objects such as chain links, bars, bolts, and various other types and shapes of metal. A cold chisel is driven by hammer blows to its flat end.

commerce. The buying and selling of goods that require transportation from one place to another.

commercial buildings. Structures designed to accommodate businesses. Commercial buildings include buildings such as stores, office complexes, and many types of community service buildings.

community college. A local, two-year school that is usually supported in part by the state and/or local government.

compactor. A compactor, or roller, is used to compact the soil of a roadway just before the road is paved. Types of compactors include steel drum rollers, tamping-foot rollers, grid or mesh rollers, and rubber-tired rollers.

com-ply. Com-ply, or composit-ply, is made of several plies of veneer strips laminated to a core of particle board.

compressive strength. A term which means something, such as a concrete driveway, can carry a lot of weight per square inch (psi).

computer-aided design (CAD). The use of computers for designing and engineering construction projects speeds up the work. CAD also provides great accuracy in design calculations.

concrete. A mixture of sand, rocks, and a binder. Concrete is one of the most common construction materials.

concrete pump. Moves concrete from the concrete mixer to the concrete form efficiently. The pump is usually mounted on a truck. It has a boom, or long arm, that can be pointed in any direction. The boom holds and directs a hose through which the concrete is pumped.

conduit. A pipe through which individual wires can be pulled. It is used in concrete walls and floors to allow easy access to the wires for maintenance and repair.

construction. The building of structures, to provide us with shelter and with places to work. Construction is also the process by which we build roads, highways, bridges, and tunnels to use in transporting people and products from place to place.

construction laser. A versatile instrument that can be used as a level or as an alignment tool. It flashes a narrow, accurate beam of light that workers can use as a baseline for additional measurements.

construction managers. People who organize all the necessary materials and people and assign the work to the workers. Managers must control and check on the work that is being done to see that it is done properly and according to specifications.

construction process. Everything that happens from the decision to build a structure to the owner's acceptance of the completed structure.

Construction Specification Institute (CSI). CSI guidelines are presented in outline form. Every type of construction job is classified in one of 16 categories. The categories follow the actual order of construction as far as possible. They also keep all the information about a subject in one place. Most specifications for large construction projects follow the format presented by the CSI.

construction superintendent. The person who controls all activity at the construction site. He or she must be aware of any problems and make corrections when they are needed. At the same time, he or she must keep a close watch on the materials that are bought and how much they cost. The cost of materials is checked through an accounting system.

construction technology. The use of tools, materials, and processes to build structures such as buildings, highways, and dams. Construction technology also relates to the knowledge we have gained about how to build structures to meet our needs.

consultant. An expert in a specific area who may be called upon to give advice on a certain part of the design of the construction project. The consultant works as a subcontractor for part of the designing and engineering work.

contact cement. An adhesive that is applied to the surfaces of materials and then allowed to dry before the materials are combined. When the materials are finally combined, the contact cement instantly forms a permanent bond.

contract. A written agreement between two or more parties. The responsibilities and rights of each party are stated in the contract. The contract contains information about the amount of work to be done, the price to be paid, and the method of payment. All parties sign the contracts. The contract is a legal document.

conveyor. Used in construction to speed the movement of materials. Roofers, for example, use conveyors to carry heavy bundles of shingles and other materials from the ground to the roof of a building.

cost-plus contract. In this type of contract, the owner agrees to pay all the costs of construction, including materials and labor. In addition, the owner agrees to pay the contractor an extra amount to cover the contractor's overhead and profit. There are two types of cost-plus contracts. One is cost plus a fixed fee. The other is cost plus a percentage of the cost.

cranes. Machines that lift large and heavy loads. They can also move loads horizontally by carrying them along a radius. Cranes are classified according to the weight they can lift safely.

crawler crane. A crane mounted on metal treads so that it can move over rough terrain at a construction site.

critical path method (CPM) chart. A chart, or diagram, made of circles and lines. Each line and circle has a meaning. This kind of schedule is useful because it shows the critical parts of the job clearly.

crosscut saw. A saw used to cut across the grain of the wood. Its teeth are shaped and sharp-ened in such a way that it actually cuts two lines very close together and removes the sawdust between them.

dam. A structure that is built across a river to block the flow of water. This is usually done for one of two reasons: to create a water reservoir for nearby communities or to collect water to power the water turbines in a hydroelectric power station.

dead load. The combined weight of all the materials in a structure. A dead load is constant; it is always there.

designing. The process of deciding what a structure will look like and how it will function.

detail drawing. A drawing that shows a particular part of the structure. It shows how things fit together. Detail drawings normally are drawn to a larger scale than other drawings.

Dictionary of Occupational Titles (DOT). A U.S. Government publication that describes over 20,000 jobs relating to many different careers.

digital rules. Measurement tools used to measure relatively long distances, such as those in highway construction projects.

dimension lumber. Lumber that measures between 2 and 5 inches thick.

drawings. Drawings show the plans for a structure in graphic form. Each of the different types of drawings shows special information that is needed for the project. Symbols are used on the drawings to represent the methods of construction and materials to be used.

ecology. The study of the way plants and animals exist together. It also studies the relationship of plants and animals to their environment.

electrical plans. These plans show the location of all the light fixtures, switches, and other electrical devices.

electricians. Workers who assemble, install, and maintain electrical wiring and fixtures.

elevations. Drawings that show the outside of the structure. An elevation drawing is made for each side of the structure.

engineering. The process of figuring out how the structure will be built and what structural materials will be used.

engineers. People who are responsible for the structural design of a project.

environment. Our surroundings, most often thought of as trees, lakes, and mountains. However, structures such as buildings, bridges, and highways are part of the environment, too.

environmental impact study. A study meant to bring to light the effect of a construction project on the environment.

equipment. A term that refers to large, complex tools and machines. Each type of equipment is designed to do a certain job. Most of the equipment that is used in construction falls into one of the following categories: surveying equipment, pumps, conveyors, and welding machines.

estimator. The person who calculates what the job will cost.

excavating. Excavating, or digging, to condition the earth for construction. This step can begin as soon as the site has been laid out.

excavator. A machine that is used for digging. It scoops up earth from one place and deposits it in another. Excavators are among the most common types of construction equipment because almost every construction job requires some excavation.

feasibility study. Research done to gather information about a proposed project. A feasibility study helps the decision-makers decide if it is feasible, or practical, to build the project. It includes information on the cost of the project and financing options as well as the availability of land and essential materials.

fiberboard. Building material made from vegetable fibers, such as corn or sugarcane stalks. It is not very strong, but it has good insulating properties. Therefore, it is used as insulation sheathing beneath the exterior siding of buildings.

final inspection. An inspection made after a construction project has been completed. The building is inspected to make sure that the job was done properly and according to the terms of the contract.

final payment. This is the last step in transferring ownership. The owner pays the construction company all the money due according to the contract.

financing. The term used to describe the process of obtaining the money used to pay for a project.

finishing stage. After the walls have been covered, the installation of the utilities is finished. This finishing stage readies the utilities for use by people.

first aid. The immediate care given to a person who has been injured.

floating. The process of moving coarse aggregate down into the concrete, leaving only fine aggregate and sand on top. This is accomplished by moving a wooden or magnesium float back and forth over the surface.

flooring. Floor covering used in buildings.

floor plan. This type of plan shows the layout of all the rooms on one floor of a building. It also shows the locations of all the walls and other built-in items. A separate floor plan is made for each floor in a building.

folding rules. Common measurement tools used to measure boards, pipe, wire, and other construction materials.

footing. The part of the structure that distributes the structure's weight. It is usually made of reinforced concrete.

forms. The molds that contain the concrete until it hardens. They are made in the shape that the finished concrete should have.

foundation. The part of the structure that is beneath the first floor. It includes the footing and the foundation walls.

foundation wall. The wall that is built directly on the footing. It transmits the weight of the superstructure to the footing.

frame structure. A frame which supports the weight of the building. A frame structure is made up of many connected frame members that are covered by sheathing. The frame members carry the weight of the structure and its contents. The framing can be made of wood, steel, or reinforced concrete.

framing square. A tool made of a single piece of steel and marked with standard or metric units. It is used to measure 90-degree angles at the corners of framework and joints. It can also be used to measure cutting angles on dimension lumber.

front-end loaders. Machines with large scoops used for shoveling. These machines are used to scoop up and deposit dirt or other materials. Loaders are often used to load trucks. Because a loader is mounted on a truck frame or a crawler, it can move small amounts of earth over short distances. Loaders have many other uses as well.

general contractor. The general contractor is the contractor in charge of the construction work.

general education. The basic courses, such as reading, writing, mathematics, science, and history, that are required in school. General education is the foundation for further learning.

geotextiles. Also called engineering fabrics. Geotextile material is like a large piece of plastic cloth. This fabric can be spread on the ground as an underlayment, or bottom layer, for roadbeds or slopes along a highway. It is used to keep soil in place and to prevent erosion.

grader. An earthworking machine that is used to grade, or level, the ground. It is used to prepare roadways and parking lots for paving.

hacksaw. A saw that is used to cut metal. Various types of hacksaw blades enable this saw to cut many different kinds of metal.

hand tools. Tools that use power supplied by a person.

hardboard. Hardboard is made up of very small, threadlike fibers of wood that are pressed together. Because the fibers are so small, when they are pressed together they form a very smooth and hard material. The fibers are held together by lignin, a natural adhesive found in the fibers.

hardwood. Wood that comes from deciduous trees. These trees shed their leaves each season.

heavy equipment. Large and powerful machines designed to do jobs that might be impossible to do by hand, such as lifting and moving earth or other heavy materials. Heavy equipment includes machines such as cranes, excavators, bulldozers, and loaders, as well as equipment that is used in highway construction.

highway construction. A general term used for the construction of any road or street. The basic steps of highway construction are preparing the soil, preparing the roadbed, and striping the finished road.

incentive contract. This type of contract is designed to reward or penalize the contractor, depending on when the job is completed. If the job is finished before the agreed-upon date, the contractor is rewarded with an amount of money that is specified in the contract. If the job is not done by the specified date, the contractor is penalized a certain amount of money.

industrial buildings. Structures which house the complex machinery that is used to manufacture goods. Industrial buildings are generally low buildings of only one or two stories.

infrastructure. The mechanical and electrical setups of a structure.

insulation. Material used to keep heat from penetrating a building in summer and cold from penetrating in winter. Insulation is usually made from spun glass, foamed plastics, or certain vegetable and mineral fibers.

interest. The price the borrower pays for using a lending institution's money.

ironworkers. The tradespeople who build steel-framed structures.

job. A paid position at a specific place or setting.

laborers. The workers who do the supportive physical work at the construction site.

labor unions. Worker-controlled organizations that are formed to present the demands of the workers to the management of construction and other types of companies.

laminated beams. Long, thin strips of wood that have been glued together. If the beam is to be curved, the wood strips are bent around a form and clamped until the glue dries. This process is repeated until the desired shape and thickness are reached. The resulting beam is strong and durable.

laminated joists. Laminated joists are lighter than dimension-lumber joists, but just as strong. They do not warp or shrink as easily as dimension lumber. They are made of three parts: two flanges and a web. The flanges are at right angles to the web, forming a cross-section that looks like the capital letter *I*. Each part is made of several plies of wood.

laser-powered welder. A welding machine used in special situations. A concentrated laser beam can heat metal to temperatures over 10,000 degrees Fahrenheit (5,540 °C). This makes the laser an ideal heat source for welding hard-to-melt metals, such as heat-resistant types of steel.

laying out. The process of identifying the location of the proposed structure on the building site. The corners and edges of the building to be constructed are marked with stakes and string. Proposed parking lots and roadways are also marked at this time.

letter of commitment. This document states the terms and conditions that are required for payment of a loan. The letter of commitment states the loan's interest rate and the number of payments in which the borrower must pay back the loan.

level. A long, straight tool that contains one or more vials of liquid. It is used to make sure that something is exactly horizontal (level), or vertical (plumb).

lien. This is similar to a claim, except that the owner's property becomes security for the amount due the worker.

live load. A variable, or changeable load. It is one that is there only some of the time.

load-bearing ability. This is the amount of weight that soil can safely support without shifting.

lump-sum contract. A contract in which a lump sum (fixed price) is paid for the work to be done. The fixed price is agreed upon before the work begins. The sum may be paid in several payments. The final payment is made when the work is completed satisfactorily.

maintenance. Taking care of an object or structure so that it continues to function in the way it was intended.

masonry. The process of using mortar to join bricks, blocks, or other units of construction. These materials are used mainly for exterior walls on houses and buildings.

masonry cement. A commercially prepared mixture of portland cement and hydrated lime.

mass structures. Structures that use solid material, such as concrete, for the building's walls.

materials. The substances from which products are made.

mechanical plans. Mechanical plans are prepared for the plumbing and piping systems.

mesh. A kind of reinforcing for concrete. Made from steel wire, mesh looks like a wire fence.

micro-lam. Made of pieces of veneer that have been laminated in a parallel direction. Thin, dried veneer is coated with waterproof adhesive and bonded under heat and pressure. Micro-lam is up to 30 percent stronger than comparable lumber. In addition, it uses 35 percent more of each tree than does regular lumber. The micro-lam process virtually eliminates warping, twisting, and shrinking.

miter gage. An attachment for the table saw. The miter gage can be adjusted to guide wood through the saw at an angle of up to 30 degrees.

modular construction. In this method of construction, a building is designed to be constructed with modules.

module. A standard unit that has been chosen by a manufacturer. Modules allow the material supplier to stock pieces in standard sizes. When the building is to be constructed, the contractor orders all the parts, which have already been cut to size in the factory. The builder needs only to assemble the pieces. Little or no cutting is needed at the site.

molding. Special trim used to cover joints where floors, walls, and ceilings meet.

mortgage note. The mortgage note is actually two documents: the note, in which the lender agrees to finance a project under certain conditions and at a specified interest rate; and the mortgage, which pledges the property as security for the loan.

mortar. A combination of masonry cement, sand, and water.

nailers. Sometimes called "nail guns" because they "shoot" nails. Most nailers are pneumatically powered.

nail set. A tool used to drive finishing nails below the surface of wooden trim and molding.

negotiate. To discuss the terms of a contract. When all parties agree on the terms, the contract is signed.

nominal size. The size of lumber when it is cut from the log.

nonferrous metals. Metals that do not contain iron.

nonmetallic sheathed cable. Cable made of several wires wrapped together inside a plastic coating of insulation. It is very flexible and easy to install.

notice of completion. This legal document lets everyone know that the job is finished.

Occupational Health and Safety Administration (OSHA). OSHA sets standards that regulate safety at construction sites.

Occupational Outlook Handbook. This handbook published by the U.S. Government contains information about 200 careers. It refers to careers as occupations.

office personnel. A general term for all of the clerical and secretarial employees of a company, together with their managers and supervisors.

on contract. To build on contract means that you have a buyer before you start. You and the buyer, or owner, sign a contract. You build the project the way the owner wants it built.

on-the-job training. Training that a person receives after he or she has been hired.

operating engineers. The people who run construction equipment.

oriented-strand board. A product made from small, crooked trees that otherwise would be unprofitable to harvest. About 60 percent of the product is made from pine. Hemlock and poplar make up the remaining 40 percent. All three woods are mixed together to make panels. Oriented-strand board can be used almost anywhere that plywood can be used.

overhead. The cost of doing business. Costs for electricity, water, telephone service, office salaries, and postage are examples of overhead costs. Other overhead costs include the costs of advertising, insurance, and office rent.

paneling. The term used to describe hardboard or plywood panels that have been prefinished. Paneling is used as a decorative finish on interior walls. It is available in a wide variety of colors, patterns, and wood grains.

particleboard. A material made of small wood chips that have been pressed and glued together. Particleboard is commonly used in houses as an underlayment between the subfloor and the floor.

pavers. Machines used in the construction of highways, parking lots, and airports. These machines place, spread, and finish concrete or asphalt paving material. Paving can be done very quickly using these machines.

payment bond. This type of bond guarantees that the contractor will pay his or her employees, subcontractors, and suppliers. This kind of bond is important because if someone is not paid, he or she can file a legal claim against the owner. If the contractor fails to pay a subcontractor, for example, the bonding company makes the payment. Thus, the owner is protected.

performance bond. This type of bond guarantees that the contractor will build the project according to the agreement. If the contractor is unable to finish the job, the bonding company is responsible for seeing that the rest of the work is done. This kind of bond guarantees that the owner will not have to pay additional money to another contractor to have the job completed.

Phillips screwdriver. A screwdriver with a tip shaped like an X. It is used to turn Phillips-head screws. Because it grips the screw better, there is less chance of slipping. This reduces the chances of damaging the screwdriver, the screw, and the object being worked on.

pipefitters. Workers who build and repair pressurized pipes to carry compressed air and steam in HVAC systems and for special applications in factories. They also build special pipe systems for transportation and for power plants.

pipe wrenches. Tools used to turn objects that are round, such as pipes. The most commonly used pipe wrench is the Stilson wrench.

plumbers. People who install, maintain, and repair the plumbing systems that carry fresh water to buildings and wastewater away from buildings. They also build drainage pipes and gas systems in commercial and industrial buildings and in homes.

plywood. One of the most commonly used wood composites. Plywood gets its name from its construction. It is made of several thin plies, or veneers, of wood that have been glued together. Each veneer is glued so that its grain is at right angles to the grain of the previous veneer. The cross-layered grains make plywood very stable and strong.

pneumatic hammers. Tools that strike with great force. Pneumatic hammers, or jackhammers, are used to break up concrete or asphalt paving.

portable circular saw. A portable power tool used to cut materials that are difficult to cut with stationary tools.

portland cement. The binder for concrete. Portland cement is a mixture of clay and limestone that has been roasted in a special oven called a kiln.

powder-actuated stud driver. A tool powered by a 22-caliber cartridge that contains gunpowder. A special nail or fastener is put into the barrel, and the tool is loaded with a cartridge. The gun is pressed against the parts to be fastened, and the trigger is pulled. The powder-actuated stud driver can drive ½-inch- to 3-inch-long pins into wood, steel, or concrete.

power drills. Tools used for drilling holes in wood, metal, and concrete. The size of a drill is determined by the chuck size and the power of the motor.

power miter saw. A circular saw mounted over a small table. The saw pivots to enable the worker to cut various angles in wood. A power miter saw is used to cut precise angles in wooden molding and trim.

power of eminent domain. The right of the government to buy property for public purposes even though the owner does not want to sell. The government condemns, or takes, the property in the interest of the general public. The owner is given a fair price for the land.

power screwdriver. Used to install and remove screws. It is similar to an electric drill. However, instead of a twist drill bit, a power screwdriver has a special screwdriver bit. The screwdriver bit fits into the head of the screw just like a screwdriver.

prefabricated units. Components of structures and even whole structures which are built in factories and shipped to the building site. They include the trim, plumbing, insulation, doors, and even molded-plastic bathrooms. The units are assembled at the construction site. This method is a quick and efficient way to construct a building.

preliminary designs. First sketches of what a structure might look like. During this process, many sketches are made and saved for evaluation at a later time.

presentation model. A model that is made to show people how the finished design will look.

private project. A construction project which belongs to an individual or to a company. Some private projects are planned to meet the needs of a company or business.

project accounting. An accounting of the progress made at the job site. To keep track of the progress, the contractor must keep accurate records of what has been done.

project control. The process of giving directions and making sure the job is done properly and on time.

project manager. The person appointed to coordinate the money, workers, equipment, and materials for the job. He or she develops a schedule and plans for the storage of goods at the job site.

pry bars. A tool used by carpenters to pry the boards used to form concrete away from the concrete after it has set. Pry bars come in many different styles.

public projects. Projects which belong to the whole community and which are planned to meet the needs of the community. These projects often are paid for with tax money.

punch list. A list of things that need to be corrected.

radial arm saw. A power tool that consists of a motor-driven saw blade that is hung on an arm over a table. This type of saw is used mostly for crosscutting and for cutting angles. A radial arm saw is usually considered a stationary power tool. It is set up at one place on a construction site. The lumber is then brought there to be cut.

reinforcing bars. Called **re-bars** for short, these are steel bars that run through the inside of the concrete. Most reinforcing bars have ridges on them that help the concrete grip the bar.

release of claims. A legal document in which the contractor gives up the right to file any claim against the owner. The contractor must check to be sure all claims have been satisfied before he or she signs the release of claims.

release of liens. A formal, written statement that everyone has been properly paid by the contractor. The release provides assurance that no liens for unpaid bills can be put on the owner's property.

renovation. The process of restoring the original charm or style of a building, while at the same time adding modern conveniences.

repair. The process of restoring an object or structure to its original appearance or working order.

residential buildings. Buildings in which people reside, or live. The two basic kinds of residential buildings are single-family units and multiple-family units.

ripsaw. A saw with chisel-like teeth designed for ripping, or cutting with the grain of the wood.

roof truss. A preassembled frame of wood or steel that is designed to support a roof. Roof trusses are lighter than other types of roof supports, but they are just as strong.

rotary hammer. A rotary hammer operates with both rotating and reciprocating action. It is used to drill holes in concrete.

roughing in. The installation of the basic pipes and wiring that must be placed within the walls, floor, and roof. This must be done before the interior walls are covered.

saber saw. This saw has a small knife-shaped blade that reciprocates (moves up and down) to cut curves. Plumbers and carpenters use saber saws to cut holes in floors and roofs for pipes.

safety factor. An extra measure of strength added to the design of a structure.

safety rules. Regulations aimed at preventing accidents and injuries in the workplace.

scale models. These models are made to help visualize how the final design will look. They have proportions that the finished building will have. However, they are built on a much smaller scale. There are two basic types: presentation models and study models.

scheduling. Estimating the amount of time it will take to do each part of the job. The schedule identifies who will do what job and in what order.

scraper. A machine that is used for loading, hauling, and dumping soil over medium to long distances.

screeding. The process of moving a straight board back and forth across the top of a form. This removes any excess concrete and levels the top of the concrete.

section drawing. A drawing that shows a section, or slice, of the structure. In this drawing, a part of the building is shown as if it were cut in two and separated. This allows the viewer to see the inside of the structure.

service drop. The service drop is the wiring that connects a building to the electric company's overhead or underground wires. A meter located near the service drop measures the amount of power used.

service panel. A box that contains circuit breakers for each individual, or branch circuit. It also contains a main circuit breaker. This controls all of the individual circuits collectively.

sheathing. Sheets of steel used to cover banks or walls of soil when the soil must be stabilized around a hole or trench.

shoring. Long strips of metal driven into the earth to hold sheathing. Shoring can also be used without sheathing to stabilize soil and reduce the chance of cave-ins.

site. The land on which a project will be constructed.

site plan. This type of plan shows what the site should look like when the job is finished. The locations of the building, new roads and parking lots, and even trees are shown. The finished contour of the earth is also shown.

slab bridge. This simple type of bridge consists of a concrete slab supported by abutments. Some of the longer slab bridges are also supported by a pier, or beam, in the middle. This type of bridge is usually made of steel or steel-reinforced concrete and is used mostly for light loads and short spans.

sledgehammers. Heavy hammers that are used to drive stakes into the ground and to break up concrete and stone.

slump test. A sample taken for testing to check the workability of the concrete.

softwood. Wood that comes from coniferous (evergreen) trees. Pine, fir, and spruce are some common softwoods that are used in construction.

specifications. These documents tell the contractor exactly what materials to use and how to use them.

specification writers. Construction company employees with the knowledge and experience needed to put specifications together. They must have a good understanding of construction practices and an up-to-date knowledge of materials. They must also be familiar with building regulations and the current costs of materials.

speculation. To build on speculation means that you build the project and then find someone to buy it.

spillway. A safety valve that allows excess water to bypass the dam. Few dams are strong enough to withstand the force of floodwater. If the water could not bypass the dam, the dam would break.

spiral ratchet screwdriver. A screwdriver that relies on a pushing force rather than a twisting force. The tips for a spiral ratchet screwdriver are interchangeable, so it can be used for Phillips-head or standard slotted screws.

standard screwdriver. A screwdriver with a flat tip designed to fit a standard slotted screw.

standard stock. Standard shapes and sizes of structural steel, such as horizontal steel beams and vertical steel columns.

staplers. This tool works like a nailer, but is loaded with U-shaped staples instead of nails. Some are pneumatic, and others use electricity. Roofers use staplers to fasten roof shingles to decking. Carpenters use them to staple insulation into place.

structural drawing. This type of drawing gives information about the location and sizes of the structural materials.

structural engineering. The process of selecting appropriate materials from which to build a structure.

structural steel. Steel that is used to support any part of a structure.

study model. This model of a design is made so that the design can be tested.

superstructure. The part of a building, above the foundation, beginning with the first floor.

surveyors. Workers who measure and record the physical features of the construction site. They survey the property and establish its official boundaries. They also lay out, or measure and mark, the construction project.

surveyor's level. A tool used to find an unknown elevation from a known one.

suspension bridges. The very longest crossings are spanned by suspension bridges. They are suspended from cables made of thousands of steel wires wound together.

table saw. The table saw consists of a blade mounted on an electric motor beneath a table-like surface. The blade sticks up through a slot in the table. The table saw is used for cutting large sheets of wood, plywood, and other wood products and for ripping lumber.

tape measures. Commonly used measurement tools especially useful for measuring long or curved surfaces.

technical institute. A technical institute is a school that offers technical training for specific careers.

technicians. Construction personnel who work in laboratories, testing soil samples and concrete and asphalt materials being used on the project.

technology. The use of technical methods to obtain practical results.

tool. An instrument of technology used to make a job easier.

tower crane. A tower crane, or climbing crane, has a built-in jack that raises the crane from floor to floor as the building is constructed.

A tower crane is used in the construction of tall buildings. It is usually positioned in the elevator shaft. When its job is done, another crane is used to remove the tower crane from the building.

transfer of ownership. A formal notice of completion, which legally establishes that the job is done.

transit. A tool that measures horizontal and vertical angles. Surveyors use it to measure relative land elevation.

trencher. A special kind of excavator that is used to dig trenches, or long, narrow ditches, for pipelines and cables. This machine is made of a series of small buckets attached to a wheel or chain. As the wheel or chain rotates, each bucket digs a small amount of earth. The operator controls the depth and length of the trench.

troweling. The final smoothing of concrete. After the concrete has begun to set, a steel trowel is used to smooth the surface.

truck crane. A crane mounted on a truck frame so that it can be driven to the site.

truss bridges. Truss bridges are supported by steel or wooden trusses, or beams that are joined to form triangular shapes. Triangular shapes are used because the triangle is a particularly strong structural shape. Trusses are also used in combination with other bridges to give them additional support.

twist drill bit. A drill with spiral grooves. A ½-inch drill will hold a bit with a diameter up to ½-inch. Some drills have reversible motors so that they can turn forward or backward. Some are even battery powered for use in places where there is no electrical service.

unit cost. The cost of the material per selling unit.

unit-price contract. In this type of contract, the contractor gives the owner a price that he or she will charge for each unit of work.

vapor barrier. A vapor barrier is placed between the inside wall of the building and the insulation to prevent water from condensing. The vapor barrier can be made of plastic, foil, or asphalt. It should always be placed on the interior side of the insulation.

waferboard. A material made of large wood chips that are pressed and glued together and then cured with heat. Waferboard is not quite as strong as plywood, but it can be used instead of plywood in many applications.

warranty. A guarantee or promise that a job has been done well or that the materials have no defects. It is usually written.

water pumps. Used to pump water out of holes in the ground so that work can be done. These pumps are usually powered by small gasoline engines. Because they do not require electricity, they are considered very portable.

wood chisels. Tools with a wedge-shaped blade, used to trim wood. They are used to pare or clear away excess material from wood joints and to remove wood to make gains (recessed areas) for hinges.

wood composites. Products that are made from a mixture of wood and other materials. Most wood composites are produced in large sheets, usually 4 feet wide and 8 feet long.

zoning laws. Cities are divided into residential, business, and industrial zones. The site for a project must be in the proper zone. Zoning laws tell what kinds of structures can be built in each zone.

||| SECTION ACTIVITIES

▐▐▐ PHOTO CREDITS

Abbey Floors, Ann Garvin, 291
Acker Drill Co., Inc., Scranton, PA, 369
Alaska Division of Tourism, 24, 35
AMP Incorporated, 350
Armstrong World Industries, Inc., 98
Arnold & Brown, 38, 51, 59, 63, 86, 285, 297
Austin Commercial, Inc./Ann Garvin, 155, 184
Autodesk Inc., 19, 147, 258

Sandar Balantini/San Francisco Convention &
    Visitors Bureau, 211
Roger B. Bean, 5, 16, 27, 29, 31, 37, 39, 45, 53, 56,
    58, 62, 82, 87, 156, 165, 183, 184, 222, 223, 225,
    232, 246, 260, 277, 278, 279, 280, 281, 282, 288,
    291, 300, 302, 344, 370
Bethlehem Steel/Bell Atlantic, 279
The Bettmann Archive, 41, 42, 263
Black & Decker, 8, 76, 116, 117, 118, 119

Canfor Limited, 91
Cardinal Industries, 140, 141
Caterpillar Inc., 126, 127, 128, 129, 130, 173, 271, 272
The Ceco Corporation, Oakbrook Terrace, IL, 274
Centimark Corporation, 123, 283
CertainTeed Corporation, 94
The Chicago Convention & Visitors Bureau, 18
CILCO, 25, 67
Coherent General, Inc., 124
Colonial Pipeline Co., 360
Combustion Engineering Inc., 43
Harold Corsini, Western Pennsylvania Conservancy,
    241
Coventry Creative Graphics, 172, 181, 186, 256
CS&A, 62, 162, 229, 307
Custom Cabinets, Carney & Sons, Ann Garvin, 86

Danforth Floor Covering, Roger B. Bean, 291
Howard Davis, 84, 88, 89, 90, 95, 96, 100, 101, 102,
    103, 110, 111, 112, 113, 114, 115, 116, 125, 182, 187,
    208, 210, 213, 214, 235, 243, 248, 252, 253, 255,
    259, 270, 273, 282, 284, 285, 286, 287, 292, 309,
    310, 311, 312, 313, 315, 316, 317, 318, 319, 320,
    321, 322, 323, 324, 325; Activities
Design Associates, Inc., 231, 237, 292
Diamond-Star Motors Corp./James Gaffney, 166

Dow Corning Corporation, 22
Duo-Fast, Beach & Barnes, 120
Duo-Fast Corporation, 104
Du Pont, 105
Dykon, Inc., 269

Emerson Electric Co., 41
English Heritage, 14

David Falconer, David R. Frazier Photolibrary Inc.,
    340
Randy Feucht Construction, Inc., Ann Garvin, 92, 275
L.B. Foster Company, 273
Fox & Jacobs, Ann Garvin, 285
Dr. Ed Francis, Dept. of Industrial Technology, Illinois
    State University, 303
Franklin Street Bridge, Roger B. Bean, 209
David R. Frazier Photolibrary Inc., 2, 21, 351, 359, 366
French Government Tourist Office, 16, 241

Jim Gaffney, 90
Ann Garvin, 2, 14, 23, 28, 30, 36, 38, 39, 44, 46, 47,
    55, 77, 80, 92, 93, 95, 97, 100, 109, 122, 145, 162,
    164, 179, 183, 198, 202, 203, 204, 209, 220, 221,
    229, 230, 233, 240, 245, 259, 267, 271, 284, 295,
    301, 343, 344, 348, 355, 357, 362, 365
Georgia Port Authority, 227
Goes Lithographing Company, Chicago, 157
Greater Peoria Airport Authority, 205
Jeff Greenberg, Ridgewood Newspapers, 49
Betty Groskin/Jeff Greenberg, 152

Bob Harr, Hedrich-Blessing/Pentair, 143
Noah Herman Sons, House of the 1990s, Roger B.
    Bean, 170, 265
Rick Herrig, Congra, Inc., 346
Hewlett Packard, 142, 156, 353, 371
Peter Honig/Jeff Greenberg Collection, 2

Idaho Transportation Department, 293
Illinois Department of Transportation, 26, 180
Illinois Laborers and Contractors Training Program,
    166
Imperial Irrigation District, Imperial, CA, 24
Ingersoll-Rand Company, 119
ITW Ramset/Red Head, 120

Chip Jamison, Pennwalt Corp., 206
Jefferson National Expansion/National Park Service, 33
Mike Jenkins, McDonald's, 289
Chris Jones, Ashland Oil Inc., 50

Kerr-McGee Corp., 203
Kruse Brothers Bldg., Inc., Roger B. Bean, 83, 96

Lab Safety Supply, Inc., 58, 60
Lake Cities Concrete, Ann Garvin, 80, 81, 275
Laser Alignment Inc., 12, 122, 139
Stef Leinwohl Photo, 57, 302
Larry Levine, 207
The Lietz Company, 121
Lindal Cedar Homes Inc., 78, 245
Michael J. London/Lone Star Industries, 178
Louisburg Construction Ltd., 131

Manhattan Construction Co./Ann Garvin, 161
The Manitowoc Company, 40, 52, 126, 200, 239, 244, 365
Manual High School, Roger B. Bean, 306, 329, 348
Master Builders, Inc., 178
McCormick Properties, Inc./Archie Johnson, 201
Robert McElwee, 158, 189, 289, 299, 360
Micro-Lam® L.U.L., 138
Herman Miller, Inc., 338
Mississippi Department of Economic Development, 215
MMFG Company, Bristol, Virginia, 137
Mason Morfit/Lone Star Industries, 135
Morton High School, Roger B. Bean, 305, 345

NASA, 134
NASA/Paul Field, 144
NASA, Kennedy Space Center, FL, 234
National Oak Flooring, 98
University of Nevada/Manis Collection, 213
New York Dept. of Economic Development, 17, 263
North Wind Picture Archives, 13, 26, 65

Paluska & Weinstein Plumbing, Roger B. Bean, 351
Peoria Public Library, Roger B. Bean, 347, 352
Judith Peterson, 61, 341

Brent Phelps, 108, 165, 219, 234, 361, 363, 372
Port Authority of New York, 202
Porter-Cable Corp., 54
The Principal Financial Group, 79, 158
Liz Purcell, 19, 82, 297
P & W Builders, Roger B. Bean, 2, 154, 169, 175, 218, 252, 266, 274, 296, 339, 349, 360, 363, 368

QA Photos Ltd., 19, 160, 199

Randolph & Associates, Architectural Services, Inc., 249, 250, 251, 252
Richwoods High School, Roger B. Bean, 352
Mark Romine, 20, 153, 238, 358

Sears Roebuck Co., 118, 119
Bob Shimer, Hedrich-Blessing/Pentair, 133
E.M. Smith & Company, Roger B. Bean, 124
R. Hamilton Smith, Master Builders, Inc., 123
Spaw-Glass General Contracting, Inc./Brent Phelps, 179
Spirit of Peoria, CILCO, 217
David S. Strickler, Strix Pix, 341
Sverdrup Corp., 257

The Tensar Corporation, 136
Tennessee Valley Authority, 32
Tennessee Wildlife Resource Agency, 43
Thompson's Food Basket, Arnold & Brown, 38
TJI®, 87

U.S. Sprint, 163

Valhi, 107

Wagner Spray Tech Corporation, 290
Walden Pond State Reservation, 327
Wallace Tile Company, Ann Garvin, 291
Doug Wilson/David R. Frazier Photolibrary Inc., 367
Wolf Creek/Colorado Tourism Board, 39
Woodruff High School, Roger B. Bean, 51, 171, 247, 326, 342, 375

Interior Design by Benoit Design and Nancy Norrall DESiGN

‖‖‖ INDEX